T0214077

Lecture Notes of the Institute for Computer Sciences, Social Informatics and Telecommunications Engineering 374

More information about this series at http://www.springer.com/series/8197

Giuseppe Caso · Luca De Nardis ·
Liljana Gavrilovska (Eds.)

Cognitive Radio-Oriented Wireless Networks

15th EAI International Conference, CrownCom 2020
Rome, Italy, November 25–26, 2020
Proceedings

 Springer

Editors
Giuseppe Caso ⓘ
Department of Mobile Systems
and Analytics (MOSAIC)
Simula Metropolitan Center for Digital
Engineering (SimulaMet)
Oslo, Norway

Luca De Nardis ⓘ
DIET Department
Sapienza University of Rome
Rome, Italy

Liljana Gavrilovska ⓘ
Faculty of Electrical Engineering
and Information Technology
Ss. Cyril and Methodius University
Skopje, North Macedonia

ISSN 1867-8211   ISSN 1867-822X   (electronic)
Lecture Notes of the Institute for Computer Sciences, Social Informatics
and Telecommunications Engineering
ISBN 978-3-030-73422-0   ISBN 978-3-030-73423-7   (eBook)
https://doi.org/10.1007/978-3-030-73423-7

This Springer imprint is published by the registered company Springer Nature Switzerland AG
The registered company address is: Gewerbestrasse 11, 6330 Cham, Switzerland

Preface

We are delighted to introduce the proceedings of the 15th edition of the EAI International Conference on Cognitive Radio Oriented Wireless Networks – CROWNCOM 2020. Despite the ongoing emergency caused by the COVID-19 pandemic, which led to the decision to switch to a fully virtual format, the conference attracted researchers from all over the world active in all fields related to cognitive radio and networks and to the role of Artificial Intelligence in this research area. The theme of CROWNCOM 2020 was in fact "Intersection and interaction between cognition and communications in the context of 5G networks and beyond".

The technical program of CROWNCOM 2020 consisted of 14 full papers presented in the main conference track, which covered all major technical aspects related to cognitive radio and networks. The presentations were organized into four sessions: Session 1, "Spectrum sensing and environment awareness", addressing physical layer issues and in particular the collection of information towards efficient coexistence; Session 2, "Resource sharing and optimization", focusing on resource sharing and network organization and optimization; Session 3, "Verticals and applications", discussing verticals enabled by cognitive radio; and finally Session 4, "Business models and spectrum management", which presented and discussed spectrum management approaches and business opportunities made possible by cognitive radio in 5G.

Aside from the high-quality technical paper presentations, the program also featured an interesting special session on the topic of "Cognitive networks in the context of software defined everything", providing new perspectives on the role of cognitive radio and networking in the next generation flexible networks and applications, highlighting their challenges and potential. We would like to thank Dr. Jorge Pereira, Principal Scientific Officer at the Future Connectivity Systems Unit of the Directorate-General Communication Networks, Content and Technology of the European Commission, for proposing and organizing this special session.

We would like to thank the chair and members of the steering committee for granting us the opportunity of organizing the 2020 edition of CROWNCOM and participating in the legacy set by this conference through the years. We would also like to thank the colleagues that made this event possible: the authors that decided to submit their work to CROWNCOM 2020, allowing us to present a high-quality program; the members of the organizing committee, the chairs and members of the Technical Program Committee (TPC); and the reviewers, fundamental for carefully selecting the best contributions. We would like to acknowledge in particular Dr. Adrian Kliks, General Co-Chair of CROWNCOM 2019 and TPC Co-Chair in the 2020 edition, for his fundamental support in all phases of the conference organization, from the definition of its scope to the management of the reviewing process. We are also grateful to Conference Managers Kristina Petrovicova and Angelika Klobusicka for their support during the organization.

We believe that the CROWNCOM conference is the perfect framework for presenting, discussing and learning about recent developments related to cognitive radio in the context of 5G and beyond 5G networks, and we are confident that future editions will continue to provide a stimulating environment for further advancement on the research topics addressed by the contributions presented in this volume and beyond.

March 2021

Giuseppe Caso
Luca De Nardis
Liljana Gavrilovska

Conference Organization

Steering Committee

Imrich Chlamtac	Bruno Kessler Professor, University of Trento, Italy
Thomas Hou	Virginia Tech, USA
Abdur Rahim Biswas	CREATE-NET, Italy
Tao Che	VTT – Technical Research Centre of Finland, Finland
Tinku Rasheed	CREATE-NET, Italy
Dominique Noguet	CEA-LETI, France

Organizing Committee

General Chair

Luca De Nardis — Sapienza University of Rome, Italy

General Co-chair

Liljana Gavrilovska — Ss. Cyril and Methodius University, Skopje, North Macedonia

Technical Program Committee Co-chairs

Jocelyn Fiorina	CentraleSupélec, France
Adrian Kliks	Poznan University of Technology, Poland

Publicity and Social Media Chair

Valentin Rakovic — Ss. Cyril and Methodius University of Skopje, North Macedonia

Publications Chair

Giuseppe Caso — Simula Metropolitan Center for Digital Engineering, Norway

Web Chair

Luca De Nardis — Sapienza University of Rome, Italy

Local Chair

Mai T. P. Le — The University of Danang - University of Science and Technology, Vietnam

Technical Program Committee

Shahwaiz Afaqui	Universitat Oberta de Catalunya, Spain
Hamed Ahmadi	University College Dublin, Ireland
Irfan Ahmed	Higher Colleges of Technology, United Arab Emirates
Özgü Alay	Simula Metropolitan Center for Digital Engineering/ University of Oslo, Norway
Marylin Arndt	Orange Labs, France
Stefan Aust	NEC Communication Systems, Ltd., Japan
Chung Shue Chen	Nokia Bell Labs, USA
Jean-Baptiste Doré	CEA-LETI, France
Serhat Erkucuk	Kadir Has University, Turkey
Stanislav Filin	NICT, Japan
Matthieu Gautier	Université de Rennes 1, IRISA, France
Andrea Giorgetti	University of Bologna, Italy
Heikki Kokkinen	Fairspectrum, Finland
Kimon Kontovasilis	National Center for Scientific Research Demokritos, Greece
Vuk Marojevic	Mississippi State University, USA
Arturas Medeisis	ITU, Lithuania
Klaus Moessner	University of Surrey, UK
Karthick Parashar	IMEC, Belgium
Milica Pejanovic Djurisic	University of Montenegro, Montenegro
Jordi Perez-Romero	Universitat Politècnica de Catalunya, Spain
Piotr Remlein	PUT, Poland
Marcin Rodziewicz	PUT, Poland
Aydin Sezgin	Ruhr-University of Bochum, Germany
Pawel Sroka	Poznan University of Technology, Poland
Victor Valls	Trinity College Dublin, Ireland
Martin Weiss	University of Pittsburgh, USA
Seppo Yrjölä	Nokia, Finland
Youping Zhao	Beijing Jiaotong University, China

Contents

Business Models and Spectrum Management

Spectrum Sensing and Environment Awareness

Active User Blind Detection Through Deep Learning

Cyrille Morin[1]($\boxtimes$) ⓘ, Diane Duchemin[1] ⓘ, Jean-Marie Gorce[2] ⓘ,
Claire Goursaud[2] ⓘ, and Leonardo S. Cardoso[2] ⓘ

[1] Univ Lyon, Inria, INSA Lyon, CITI, Lyon, France
{cyrille.morin,diane.duchemin}@inria.fr
[2] Univ Lyon, INSA Lyon, CITI, Lyon, France
{jean-marie.gorce,claire.goursaud,leonardo.cardoso}@insa-lyon.fr

Abstract. Active user detection is a standard problem that concerns many applications using random access channels in cellular or *ad hoc* networks. Despite being known for a long time, such a detection problem is complex, and standard algorithms for blind detection have to trade between high computational complexity and detection error probability. Traditional algorithms rely on various theoretical frameworks, including compressive sensing and bayesian detection, and lead to iterative algorithms, e.g. orthogonal matching pursuit (OMP). However, none of these algorithms have been proven to achieve optimal performance.

This paper proposes a deep learning based algorithm (NN-MAP) able to improve on the performance of state-of-the-art algorithm while reducing detection time, with a codebook known at training time.

Keywords: Non-coherent active user detection · Machine learning · Massive random access

1 Introduction

Massive access in internet of things (IoT)-dedicated radio networks, especially in 5G, brings several challenges. In this setting, a huge number of sensor nodes is to be sporadically served within the specific constraints of machine type communication (MTC). These networks will be implemented mainly with low cost devices, thus having restricted radio functionalities as well as scarce power and computational resources. Besides, latency, spectrum and energy, must be held to the same efficiency demands of current communication standards, sometimes even higher, to fulfil the requirements of the foreseen tactile internet [1]. Uplink data transmissions from simplified sensor nodes aim at minimising transmission duration as well as the amount of transmissions, given their finite power resources, but also the small amount of data to be transmitted. A high signalling overhead would drastically reduce the operational life-time of such devices as they would spend more time and energy to transmit protocol-related messages than useful payload data. Unlike current 4G based access procedures, as planned in

© ICST Institute for Computer Sciences, Social Informatics and Telecommunications Engineering 2021
Published by Springer Nature Switzerland AG 2021. All Rights Reserved
G. Caso et al. (Eds.): CrownCom 2020, LNICST 374, pp. 3–15, 2021.
https://doi.org/10.1007/978-3-030-73423-7_1

narrowband IoT (NB-IoT) (even though new releases provide a shorter access procedure [2]), an "all in one" grant-free uplink message encapsulating access request, device identifier, and data, would be ideal.

To achieve this grant-free uplink reality, the main challenge is the detection of the active subset of sensor nodes by the base station (BS), also referred to as active user detection (AUD). To enable the transmission of users' identities despite a high network density, the usage of a dedicated spectrum sharing technique is required that must be compatible with the rapidly evolving traffic load within a high number of potential users. Non orthogonal multiple access (NOMA) [3], and in particular code-domain NOMA, is a good candidate [4] for such a spectrum sharing technique as it limits collisions for simultaneous transmissions without requiring to use an extremely long access sequence for each user, given the network density.

As a result, all the complexity of the AUD task is pushed to the BS. Avoiding a handshake procedure requires efficient detection algorithms to retrieve active users' identity from a "one shot" access message with limited channel state information at the receiver (CSIR). The optimal AUD as described in [5] suffers from high complexity which does not seem compatible with real time implementation. An iterative version of the optimal detector, having a lower complexity -but also lower performance-, is also introduced therein. Developing a high performance though low complexity detector is crucial for a realistic and efficient AUD implementation. Most efficient algorithms proposed in the literature to cope with this problem exploit either a Bayesian estimation formalism or the compressive sensing formulation [6,7]. Both have many similarities but lead to different iterative algorithms. Despite their efficiency, none of these algorithms can guarantee to achieve the optimal solution as they have to trade their accuracy with complexity. Therefore, the competition is still open.

With the recent and growing interest of the community toward machine learning, and particularly deep learning (DL), it has been shown that its usage can help to solve complex problems, mainly when defining good models is difficult, or when the models exist but provide solutions too complex for their exploitation. The scenario presented here falls into the second category, and appear to be a good candidate to exploit DL. The objective of this paper is to design a DL receiver for massive NOMA, and more specifically, the AUD in non-coherent channels. Related studies have been done around this subject, for instance in [1,8]. These works are focused on the resource optimisation problem for code domain NOMA and employ auto-encoder based solutions in both cases. They show that the encoding and decoding performance can be improved through end-to-end optimisation. The metrics used there are symbol error rate (SER), sum rate and convergence rate. In [9], the question of imperfect CSIR is addressed for power domain NOMA. This paper is also addresses resource allocation optimisation. The authors of [10], while also dealing with power domain NOMA, focus on channel estimation and signal detection in the context of orthogonal frequency-division multiplexing (OFDM). They propose a comparison with a successive interference cancellation (SIC) based algorithm and show the interest of the DL approach. The model is nevertheless restricted to two users and the channel realisation is fixed in training and testing phases. A preamble and collision detection scheme based on DL is proposed in [11], where the study is performed on pre-processed long term evolution (LTE) random access preamble signals: the correlations with the possible Zadoff-Chu sequences are directly provided to the network. The objective of the authors include the detection of multiple collisions in order to improve contention resolution, and therefore, access probability. Whereas the collision study is realised in a massive access scenario, the detection evaluation is performed with a single user scenario only, by comparing the missed detection performance of the proposed fully connected neural network (NN) with other more classic preamble detectors. All these works are closely related to the use-case of the present paper, but none of them directly address the task at hand: to the best of our knowledge, this work is the first to apply DL on non-coherent AUD with code domain NOMA.

The rest of the paper is organised as follows: Sect. 2 presents our model and the reference schemes to which our approach is compared. Section 3 provides details on the implementation choice regarding the DL scheme we propose, while Sect. 4 is dedicated to the evaluation of the solution. Section 5 concludes the paper.

2   System Model for the Massive Random Access

2.1   Non-coherent AUD

The random access channel in massive MTC is important to guarantee a fair radio medium access. It is herein assumed that the sensor nodes, henceforth referred to simply as *nodes*, receive a random code in advance which is used to send a resource request to the BS they are associated with. In a standard approach, if two nodes request a resource in the same slot, a collision occurs and at least one of the two messages is lost. However, with the knowledge of the codes distributed to the nodes, the BS tries to determine the identity of all the nodes involved in a request. Such an approach, referred as coded random access [6], allows to reduce the number of resources reserved for the random access mechanism and can accelerate the handshake mechanism. Indeed, unlike the 4G access protocol which the NB-IoT is based on and relying on a pool of available Zadoff-Chu access sequences, this approach ensures the uniqueness of the codes employed by the nodes. This fact allows to avoid access code collisions but also additional steps, known as the contention resolution, dedicated to the identification of the users in the handshake procedure. It can also be used as a standalone mechanism in cases where the only one bit is to be transmitted, then the binary value of the nodes' activity suffices without needing further handshake.

In our model, the BS transmits a beacon allowing the nodes to be roughly synchronised and to control their power such that in average the received power at the BS is constant for all nodes. However, the instantaneous channel states are not known, and no pilots are used in this detection phase. The detector thus operates in non-coherent detection mode [5].

We adopt the following notation for the remainder of this work: $(\mathcal{U}, \Phi)$ is a measurable space where $\mathcal{U}$ denotes the total set of nodes, with cardinality $K = |\mathcal{U}|$ and $\Phi = \mathcal{P}(\mathcal{U})$ the powerset of $\mathcal{U}$. A node subset is denoted by $\mathcal{A} \in \Phi$. For a given random access slot, we note $\underline{\mathcal{A}} \in \Phi$ a set of *active* nodes. The activity rate is assumed low (less than 0.5) implying a sparse transmission set. We further assume that the node activity follows a Poisson distribution with mean parameter λ (thus the node activity probability is $\theta = \lambda/K$). As stated previously, a unique codebook $\mathcal{C}$ is generated and shared among the network (the transmitters and BS both agree on the codes during an initial association phase). As a result, each node k owns a dedicated complex Gaussian code $\mathbf{c}_k$ of codelength M and unit power.

As mentioned previously, the received messages are considered synchronous and a perfect average power control allows the messages to be received with an average signal-to-noise ratio (SNR) ρ. The BS possesses N antennas while the nodes have a single antenna. Transmissions are subject to a flat Rayleigh block fading channel, modelled as a random vector $\mathbf{h}_k \sim \mathcal{N}_{\mathscr{C}}(0, \mathbf{I}_N)$ of size N where $\mathbf{I}_N$ is the identity matrix of dimension n and $\mathcal{N}_{\mathscr{C}}(0, \cdot)$ indicates a complex standard Gaussian distribution. The receiver noise introduces an additive white Gaussian noise (AWGN), modelled as a random vector $\mathbf{z} \sim \mathcal{N}_{\mathscr{C}}(0, \mathbf{I}_{NM})$ of size NM. It should be noted that neither the BS nor the transmitting nodes are aware of the actual channel realisations, but only know the channel statistics, as described above. For a given active node k, the channel coefficients $h_{m,n}$ are constant with respect to (w.r.t.) to m and are independent and identically distributed (i.i.d.) w.r.t. n. This means that the message is sent over a narrowband channel, typically a single carrier in an OFDM frame, as defined in NB-IoT for MTC. The proposed model is similar to the one used in [5,7].

Let $\mathbf{y} \in \mathbb{C}^{NM}$ denote the received signal, ρ the targeted SNR and $\otimes$ the Kronecker product. The received signal is then given by:

$$\mathbf{y} = \sum_{k \in \underline{\mathcal{A}}} \sqrt{\rho}(\mathbf{I}_N \otimes \mathbf{c}_k)\mathbf{h}_k + \mathbf{z}. \tag{1}$$

The BS performs an AUD given $\mathbf{y}$ and prior knowledge, restricted to the codebook, the activity probability law and the statistical CSIR. The AUD algorithm is performed on a non-coherent channel, since no pilots are used for prior channel estimation. Let $\hat{\mathcal{A}}$ denote the detected active node subset. To evaluate the performance of the algorithm, the following metrics will be used: codeset error rate (CER), user error rate (UER), misdetection rate (MDR) and false alarm rate (FAR), according to the following definitions:

$$\text{MDR}: \quad \bar{\epsilon}_{md} = \mathbb{E}_k\left[\mathbb{P}[k \notin \hat{\mathcal{A}} | k \in \underline{\mathcal{A}}]\right] \tag{2}$$

$$\text{FAR}: \quad \bar{\epsilon}_{fa} = \mathbb{E}_k\left[\mathbb{P}[k \in \hat{\mathcal{A}} | k \notin \underline{\mathcal{A}}]\right] \tag{3}$$

$$\text{UER}: \quad \bar{\epsilon}_s = \bar{\epsilon}_{md} \cdot \theta + \bar{\epsilon}_{fa} \cdot (1 - \theta) \tag{4}$$

$$\text{CER}: \quad \bar{\epsilon}_C = p[\hat{\mathcal{A}} \neq \underline{\mathcal{A}}]. \tag{5}$$

The MDR (resp. FAR) corresponds to the false negative (resp. false positive) rate. The UER combines these errors to compute an average individual error rate. In addition, the CER is a system level error rate, that counts the rate of non-perfect codeset detection.

2.2 MAP Detectors

Let $\mathbf{y}_n \in \mathbb{C}^M$ denote the received signal on antenna n and $\mathbf{C}_\mathcal{A} \in \mathbb{C}^{M \times \omega}$ the codeset of a given node subset $\mathcal{A}$ whose cardinality is ω. Its singular value decomposition (SVD) is written $\mathbf{C}_\mathcal{A} = \mathbf{V}\boldsymbol{\Gamma}\mathbf{U}$, where $\mathbf{V} \in \mathbb{C}^{M \times M}$ and $\mathbf{U} \in \mathbb{C}^{\omega \times \omega}$ are unitary matrices. $\boldsymbol{\Gamma} \in \mathbb{C}^{M \times \omega}$ is composed of the singular values γ on its diagonal. From (2.1), following [5], the likelihood of a codeset is given by:

$$p(\mathbf{y}|\mathcal{A}) = \prod_{n=1}^{N} \frac{1}{\pi^M |\sigma|} \exp\left(\|\tilde{\mathbf{y}}_n\|_2^2 - \|\mathbf{y}_n\|_2^2 \right), \tag{6}$$

where $\sigma \in \mathbb{C}^{M \times M}$ is $\sigma = \rho \mathbf{C}_\mathcal{A} \mathbf{C}_\mathcal{A}^{\mathsf{H}} + \mathbf{I}_M$ and $\tilde{\mathbf{y}}_n$ is the projection of $\mathbf{y}_n$ onto the codeset $\mathbf{C}_\mathcal{A}$ space, and is defined as:

$$\tilde{\mathbf{y}}_n = diag\left(\sqrt{\frac{\rho|\gamma_1|^2}{1 + \rho|\gamma_1|^2}}, \cdots, \sqrt{\frac{\rho|\gamma_M|^2}{1 + \rho|\gamma_M|^2}} \right) \mathbf{V}^{\mathsf{H}} \mathbf{y}_n. \tag{7}$$

The maximum likelihood estimate (MLE) has been used in [5] to estimate the active set. Since we know the prior probability on $(\mathcal{U}, \Phi)$, related to the Poisson distribution, a maximum a posteriori (MAP) detector can be defined and is optimal w.r.t. to the Bayes risk minimisation, when defined from the CER.

Definition 1 (C-MAP estimate). *The C-MAP estimate of the codeset detection problem is given by:*

$$\hat{\mathcal{A}}_C = \arg\min_{\mathcal{A} \in \Phi} p[\underline{\mathcal{A}} \neq \mathcal{A}|\mathbf{y}] \tag{8}$$

$$= \arg\max_{\mathcal{A} \in \Phi} p(\mathbf{y}|\mathcal{A})p(\mathcal{A}), \tag{9}$$

where $\hat{\mathcal{A}}_C$ is the detected subset, given the received signal $\mathbf{y}$, from the knowledge of the codebook $\mathcal{C}$, the Gaussian distributions of the channels $\mathbf{h}_k$ and noise $\mathbf{z}$.
In addition, the prior probability is given by $\mathbb{P}(\mathcal{A}) = \lambda^{|\mathcal{A}|} \cdot (1 - \lambda)^{|\mathcal{U}| - |\mathcal{A}|}$.

But if the objective is to minimize the user error rate, the Bayes risk is modified and leads to the following estimate.

Definition 2 (U-MAP estimate). *The U-MAP estimate of the codeset detection problem is given by:*

$$\hat{\mathcal{A}}_U = \cup_{k \in \mathcal{U}} \{k|\delta_k(\mathbf{y}) = 1\}, \tag{10}$$

with δ the delta function and where $\delta_k(\mathbf{y}) = 1$ is given by:

$$\delta_k(\mathbf{y}) = \begin{cases} 1 \text{ if } \sum_{\substack{\mathcal{A}\in\Phi; \\ k\in\mathcal{A}}} p(\mathbf{y}|\mathcal{A})p(\mathcal{A}) > \sum_{\substack{\mathcal{A}\in\Phi; \\ k\notin\mathcal{A}}} p(\mathbf{y}|\mathcal{A})p(\mathcal{A}) \\ 0 \text{ else} \end{cases}. \quad (11)$$

Let us prove that U-MAP is optimal with respect to the UER metric. $P_{MD}(k|\mathbf{y})$ and $P_{FA}(k|\mathbf{y})$, the probability of Missed Detection, respectively False Alarm, of a user k given a received signal $\mathbf{y}$ are given by:

$$P_{MD}(k|\mathbf{y}) = \sum_{\substack{\mathcal{A}\in\Phi; \\ k\in\mathcal{A}}} \mathbb{1}_{[k\notin\hat{\mathcal{A}}(\mathbf{y})]} p(\mathcal{A}|\mathbf{y}) = \mathbb{1}_{[k\notin\hat{\mathcal{A}}(\mathbf{y})]} \sum_{\substack{\mathcal{A}\in\Phi; \\ k\in\mathcal{A}}} p(\mathcal{A}|\mathbf{y}), \quad (12)$$

and

$$P_{FA}(k|\mathbf{y}) = \sum_{\substack{\mathcal{A}\in\Phi; \\ k\notin\mathcal{A}}} \mathbb{1}_{[k\in\hat{\mathcal{A}}(\mathbf{y})]} p(\mathcal{A}|\mathbf{y}) = \mathbb{1}_{[k\in\hat{\mathcal{A}}(\mathbf{y})]} \sum_{\substack{\mathcal{A}\in\Phi; \\ k\notin\mathcal{A}}} p(\mathcal{A}|\mathbf{y}). \quad (13)$$

Minimising the UER thus corresponds to performing a binary test for each user, by comparing $P_{MD}(k|\mathbf{y})$ and $P_{FA}(k|\mathbf{y})$:

$$\delta k(\mathbf{y}) = 1 \text{ if } \sum_{\substack{\mathcal{A}\in\Phi; \\ k\in\mathcal{A}}} p(\mathcal{A}|\mathbf{y}) > \sum_{\substack{\mathcal{A}\in\Phi; \\ k\notin\mathcal{A}}} p(\mathcal{A}|\mathbf{y}). \quad (14)$$

Then, having $p(\mathcal{A}|\mathbf{y}) \propto p(\mathbf{y}|\mathcal{A})p(\mathcal{A})$, the decision given in (11) is optimal.

To compute the estimate given either by Eq. (9) or (11), each element of the power set Φ has to be evaluated, making such kind of AUD non feasible for computational reasons.

2.3   It-MAP Detector

As an alternative to these solution with prohibitive computational complexity, many iterative algorithms have been proposed in the literature [5,7,12]. Following the work presented in [5], the Iterative-MAP (It-MAP) is herein proposed as a reference solution, built as an approximation of C-MAP. The philosophy of It-MAP is similar to that of a Successive Interference Cancellation (SIC) as it processes the received signal $\mathbf{y}$ iteratively and retrieves a new detected user at each iteration i based on the assumption of the previous $\hat{\mathcal{A}}$ at $i-1$. Even if the detected subset is built iteratively, the detection rule is based on the MAP criteria given by Eq. (9) for each i, with a search restricted on some elements of Φ. More precisely, the evaluated subsets $\mathcal{A}_i \in \Phi_i$ are built from the previously detected subset $\hat{\mathcal{A}}_{i-1}$ as follows:

$$\Phi_i = \cup_{k\in\{\mathcal{U}\setminus\hat{\mathcal{A}}_{i-1},\emptyset\}} \left\{\hat{\mathcal{A}}_{i-1}, k\right\} \quad (15)$$

The It-MAP detection stops as soon as two successive iterations provide the same detected subset, i.e., $\hat{\mathcal{A}}_{i-1} = \hat{\mathcal{A}}_i$.

The architecture of this MAP-based AUD algorithm makes its complexity lower than the computation requirement of the MAP, but at the cost of a reduced accuracy since the iterative detection makes the It-MAP prone to error propagation. This fact let room for other AUD algorithms seeking for a better complexity-accuracy trade-off. As DL is envisioned to accommodate well to large scale problems, a blind AUD based on DL is thus presented in the following section and will be compared to the C-MAP, U-MAP and It-MAP detectors.

3   A Neural Network Based Algorithm

3.1   The NN-MAP Estimate

Definition 3 (NN-MAP estimate). *A NN-MAP architecture for the AUD problem is defined as follows. The inputs to the NN-MAP detector come from* $\mathbf{y}$, *from Eq. (1), and* ρ *(in dB) as a side information. It outputs a vector* $\mathbf{p}$ *of length* K *containing the estimated probability that each node is active. That probability is obtained by using a* sigmoid *activation function at the end of the network and is compared with a vector of ground truth labels* $\mathbf{t}$ *of the same length through a binary cross-entropy loss function* $\mathcal{L}$ *to optimise the network's parameters:*

$$\mathcal{L}(\mathbf{t}, \mathbf{p}) = -\sum_{k=1}^{K} t_k \cdot \log\left(p_k\right) + (1 - t_k) \cdot \log\left(1 - p_k\right) \tag{16}$$

During the training phase, the average cost over all tuples $(\underline{\mathcal{A}}_i, \mathbf{y}_i, \mathbf{p}_i)$ *is:*

$$\bar{\mathcal{L}} = \frac{1}{I} \sum_i \mathcal{L}(\mathbf{t}(\underline{\mathcal{A}}_i), \mathbf{p}_i), \tag{17}$$

where I *is the mini-batch size.*

After training, the soft probabilities are converted to hard decisions using a threshold: a user k *is considered active if* $p_k > 0.5$.

This choice is justified by the following theorem.

Theorem 1. *For a non-coherent AUD problem, the solution that minimizes the cost function given by* (17) *converges to the U-MAP estimate of Definition 2 if the dataset is large enough.*

Proof. By incorporating (16) into (17), and permuting the sums w.r.t. i and k, one can write:

$$\bar{\mathcal{L}} = \sum_{k \in \mathcal{U}} \bar{\mathcal{L}}(k), \tag{18}$$

with :

$$\bar{\mathcal{L}}(k) = -\sum_{i=1}^{I} \mathbb{1}_{[k \in \underline{\mathcal{A}}_i]} \log\left(p_k(\mathbf{y}_i)\right) + \mathbb{1}_{[k \notin \underline{\mathcal{A}}_i]} \log\left(1 - p_k(\mathbf{y}_i)\right). \tag{19}$$

Then if the tests are randomly selected according to the prior probability $\mathbb{P}(\underline{A})$, one have:

$$\bar{\mathcal{L}}^*(k) = \lim_{I \to \infty} \bar{\mathcal{L}}(k) \tag{20}$$

$$= -\mathbb{E}_{\underline{A},Y} \left[\mathbb{1}_{[k \in \underline{A}]} \log (p_k(\mathbf{y})) + \mathbb{1}_{[k \notin \underline{A}]} \log (1 - p_k(\mathbf{y})) \right]. \tag{21}$$

Finally, using the decomposition $\bar{\mathcal{L}}^*(k) = \int_{\mathcal{Y}} \bar{\mathcal{L}}^*(k|\mathbf{y}) \cdot f_Y(\mathbf{y}) \cdot d\mathbf{y}$, one gets:

$$\bar{\mathcal{L}}^*(k|\mathbf{y}) = -P(k|\mathbf{y}) \cdot \log (p_k(\mathbf{y})) + (1 - P(k|\mathbf{y})) \cdot \log (1 - p_k(\mathbf{y})). \tag{22}$$

The minimum of (17) is asymptotically achieved if the NN returns for each observation $\mathbf{y}$, the output $\mathbf{p}$ which minimizes (22) for any user k, independently.

Thanks to the convexity of this function w.r.t. $p_k(y)$ it is straightforward to show that the global cost function is minimal when $p_k(\mathbf{y}) = P(k|\mathbf{y})$, which is nothing but the posterior probability. Therefore, if the learning phase succeeds to find a set of parameters for the NN such that the output probability vector converges to the posterior probability distribution, the U-MAP estimate is achieved by selecting at the output all the nodes k with $p_k \geq 0.5$. Clearly, the NN architecture approximates the U-MAP.

It is worth mentioning that there is no guarantee that a NN can achieve this global optimum. This theorem only claims that the objective function based on the cross-entropy is well-posed w.r.t. the U-MAP problem.

3.2   The NN-MAP System Parameters

In this paper, the NN architecture and hyper parameters have been empirically optimised as follows, for a scenario chosen with $K = 10$, $M = 8$, $N = 4$, $\lambda = 4$. One random codebook is generated and reused for all subsequent training for a fair comparison of the performance thus obtained. Unless specified otherwise, all figures correspond to this scenario.

Architecture Type: The observation vector $\mathbf{y}$ is a combination of random Gaussian variables related to the properties of the codebook, channel realisations and noise. The unique correlation may come from the codebook, which is selected randomly. As such, the correlations are low and convolutional layers are not mandatory. In addition, we don't assume in this model any correlation in the data transmissions, neither between nodes, nor over time. Under these assumptions, a recurrent layer is not necessary. Consequently, the chosen architecture is a fully-connected neural network.

Layers: A set of networks was trained with an increasing amount of dense layers and a constant amount of units in each one, from 3 to 12. The search was not expanded above due to lack of significant improvement. Finally, the 5 layers network was selected.

Units: The number of units per layer was set to be proportional to K, M, and N so it can scale according to scenario complexity. The proportionality factor was chosen by increasing it by powers of 2 from 1 to 128. Performance stopped increasing after 4, so this is what is used in the following.

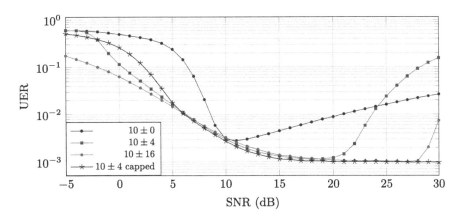

Fig. 1. Network trained on examples with uniform SNR. Mean is kept at 10 dB and range is varied. For the capped line, the SNR value input to the NN at test time is limited to the training range.

SNR Range: Training data is generated with a given ρ, which is also input to the NN. It is therefore necessary to determine the values that will be used to train the network with. Indeed, ρ can have a big impact on the overall performance of the trained network: a too low value could generate a very noisy signal from which the NN would fail to learn anything. A too high SNR would not help the NN to learn to cope with noise. In our approach, the training dataset is generated with a range of SNR values, distributed uniformly over a specified interval. To find out a good interval, the impact of two parameters is evaluated: the range of the interval (Fig. 1) and the mean of that interval. In those two cases, NN shows good performance only inside the SNR interval used for training, with a rapid decrease outside.

Note that when a NN has to work above its learned range, it appears more efficient to limit the SNR values input to the range bounds. This is highlighted in the curve ($—*—$), where the NN learned in the range 10 ± 4 and performs well at SNR above 14 dB. This is not the case for low SNR values, where the NN learned in the range 10 ± 16 outperforms all other curves.

The most important parameters of the NN used in this paper are summarized in Table 1.

Table 1. Summary of network and training parameters

Parameter	Value
Layers	5
Units	$4 \times K \times M \times N$
Learning rate	1×10^{-3}
Optimiser	Adam
Batch size	4096
Training iterations	100000
Training SNR values	10 dB $\pm$ 16 dB

4   Results

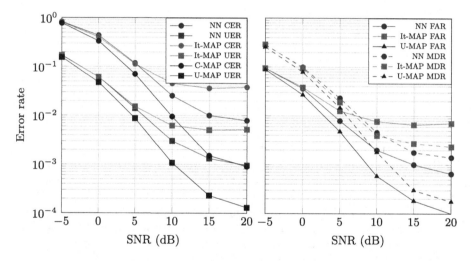

Fig. 2. Performance comparison between C-, U-, It- and NN-MAP algorithms for code-set and user error rates.

Comparison Between the Different Algorithms: The different algorithms defined in the previous section are compared for the chosen scenario ($K = 10$, $M = 8$, $N = 4$, $\lambda = 4$).

As shown in Fig. 2, the trained NN-MAP outperforms It-MAP, especially at high SNR w.r.t. UER and CER metrics as well. The performance of NN-MAP bridges half of the gap with the MAP's optimal given either with C-MAP or U-MAP. In addition, as can be seen in Fig. 2, NN-MAP outperforms It-MAP for both MDR and FAR criteria. It is worth noting however that It-MAP achieves a MDR lower than FAR while it is the opposite for NN-MAP. Note that this trade-off may be easily tuned with NN-MAP. Indeed, in (22), we proved the

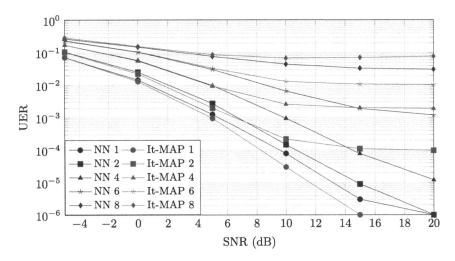

Fig. 3. It-MAP and NN-MAP are set and trained with $\lambda = 4$. A test time, active user number is not random and varied from 1 to 8.

relationship between the MAP and the loss function. It is known that the MAP solution is optimal only if mis-detection and false alarm errors have the same cost. However, it is known that the relative weight of these errors can be tuned to achieve any point of the ROC curve of the detector [13]. For the NN-MAP, one have two options to balance MDR and FAR: the cross-entropy can be modified or the hard decision threshold can be tuned.

All algorithms tested in this paper assume the knowledge of λ as a prior information. Figure 3 shows how It-MAP and NN-MAP perform when the actual number of users deviates from the expected value (the actual number of active users is indicated in the legend). It-MAP outperforms NN-MAP only for 1 active user. It is interesting to mention that a method to increase the performance of NN-MAP in that regard could be to either train it with several values of λ, or to use a uniform distribution instead of the distribution associated to the Poisson distribution assumption.

Analysis on Larger Scenarios: In Fig. 4, the performance results of It-MAP and NN-MAP are given when $K = 20$. Note that in this scenario, both U-MAP and C-MAP are not computable in reasonable time. There, NN-MAP outperforms It-MAP, since it provides a gain in terms of UER without compromising on the CER.

Computational Considerations: On top of performance increases, the NN approach also provides a reduction of computation times as shown in Table 2, where CPU executions are single thread and the GPU execution allows a batch size of 10000. The source of this reduction is twofold: a less complex computation through simple add and multiply operations, without branching leads to a reduced load and smoother execution on the system, and it also creates the

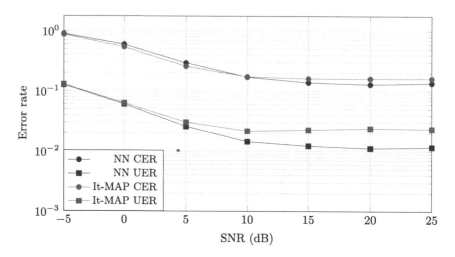

Fig. 4. Performance comparison between NN-MAP and It-MAP algorithms for codeset and user error rates on a scenario with more potential users: $K = 20$.

capability to massively parallelise batches of computations in an efficient manner. The second aspect is most prevalent in the present case: the It-MAP rests on sequential operations with nested loops that create inefficiencies through CPU branch prediction misses and are not trivial to convert to parallel computation, and so, to offload to high performance accelerators such as a GPU.

Table 2. Average example processing time for a scenario with $K = 10$, $M = 8$, $N = 4$, $\lambda = 4$.

MAP CPU	It-MAP CPU	NN CPU	NN GPU
159 ms	8.25 ms	3.6 ms	542 ns

5   Conclusion

In this paper, we have proposed to use DL for the non-coherent AUD problem based on coded domain NOMA. We have proposed a NN-MAP detector which minimises asymptotically the MAP cost function. We have shown that this approach improves on the performance of state-of-the-art iterative algorithms by a factor of 5 in some scenarios, especially w.r.t. the UER metric. Moreover, as it also reduces the algorithmic complexity, this work shows the interest of using DL for such a task even though more work still needs to be done to improve the scalability of the architecture to very large sets of nodes as expected for massive access IoT.

Acknowledgement. This work has been supported by the French National Agency for Research (ANR) under grant no ANR-16-CE25-0002 - EPHYL and Nokia Bell Labs in the framework of the Inria-Nokia Bell Labs common lab.

References

1. Ye, N., Li, X., Yu, H., Wang, A., Liu, W., Hou, X.: Deep learning aided grant-free NOMA toward reliable low-latency access in tactile internet of things. IEEE Trans. Ind. Inf. **15**(5), 2995–3005 (2019)
2. 3GPP, Further NB-IoT enhancements (RP-171428) (2017)
3. Ding, Z., Lei, X., Karagiannidis, G.K., Schober, R., Yuan, J., Bhargava, V.K.: A survey on non-orthogonal multiple access for 5G networks: research challenges and future trends. IEEE J. Sel. Areas Commun. **35**(10), 2181–2195 (2017)
4. Paolini, E., Stefanovic, C., Liva, G., Popovski, P.: Coded random access: applying codes on graphs to design random access protocols. IEEE Commun. Mag. **53**(6), 144–150 (2015)
5. Duchemin, D., Chetot, L., Gorce, J., Goursaud, C.: Coded random access for massive MTC under statistical channel knowledge. In: IEEE 20th International Workshop on Signal Processing Advances in Wireless Communications (SPAWC), 2019, pp. 1–5 (2019)
6. Wunder, G., Stefanović, Č., Popovski, P., Thiele, L.: Compressive coded random access for massive MTC traffic in 5G systems. In: 2015 49th Asilomar Conference on Signals, Systems and Computers, pp. 13–17. IEEE (2015)
7. Ke, M., Gao, Z., Wu, Y., Gao, X., Schober, R.: Compressive sensing-based adaptive active user detection and channel estimation: massive access meets massive MIMO. IEEE Trans. Signal Process. **68**, 764–779 (2020)
8. Kim, M., Kim, N., Lee, W., Cho, D.: Deep learning-aided SCMA. IEEE Commun. Lett. **22**(4), 720–723 (2018)
9. Liu, M., Song, T., Gui, G.: Deep cognitive perspective: resource allocation for NOMA-based heterogeneous IoT with imperfect SIC. IEEE Internet Things J. **6**(2), 2885–2894 (2019)
10. Narengerile, Thompson, J.: Deep learning for signal detection in nonorthogonal multiple access wireless systems. In: 2019 UK/China Emerging Technologies (UCET), pp. 1–4, August 2019. https://doi.org/10.1109/UCET.2019.8881888. ISSN null
11. Magrin, D., Pielli, C., Stefanovic, C., Zorzi, M.: Enabling LTE RACH collision multiplicity detection via machine learning (2018). arXiv: 1805.11482 [cs.IT]
12. Cirik, A.C., Mysore Balasubramanya, N., Lampe, L.: Multi-user detection using ADMM-based compressive sensing for uplink grant-free NOMA. IEEE Wirel. Commun. Lett. **7**(1), 46–49 (2018)
13. Poor, H.V.: An Introduction to Signal Detection and Estimation. Springer, New York (2013). https://doi.org/10.1007/978-1-4757-2341-0

Spectrum Sensing Based on Dynamic Primary User with Additive Laplacian Noise in Cognitive Radio

Khushboo Sinha(ID) and Yogesh N. Trivedi$^{(\boxtimes)}$(ID)

Nirma University, Ahmedabad 382470, Gujarat, India
{18ftphde24,yogesh.trivedi}@nirmauni.ac.in

Abstract. In this paper, spectrum sensing techniques with dynamic primary user (PU) are considered in the environment of Laplacian noise. It means the PU may not be present or absent during the whole sensing time. However, PU arrives or departs randomly in the sensing time interval. We consider three different detection schemes such as energy detection, absolute value cumulation detection (AVCD) and improved AVCD (i-AVCD). We present the performance in terms of receiver operating characteristic (ROC) and detection probability versus average signal-to-noise ratio (SNR) using simulations. We conclude that the detection performance in the dynamic scenario is better than the performance in the static scenario, when the arrival/departure parameter ($\theta_A T / \theta_D T$) is beyond one, where θ_A and θ_D are corresponding to arrival rate and departure rate of the PU respectively, and T is the sampling interval. Furthermore, the i-AVCD scheme outperforms AVCD and energy detection in the considered scenario.

Keywords: Spectrum sensing · Cognitive radio · Dynamic primary user · Detection probability · False alarm probability

1 Introduction

In the today's scenario of 5G communication, scarcity of microwave spectrum is the biggest bottleneck for introducing real time services. In this case, use of millimeter (mm) wave can provide enough spectrum but it demands for drastic change in the Radio Access Network (RAN), Air interface, physical layer and MAC Layer of cellular networks. On the other side, it has been observed that the microwave spectrum is not utilized effectively. Major part of this spectrum has been found vacant, we call it as spectrum holes. Thus, the spectrum holes indicate segments of the spectrum which are not utilized by the licensed users, also known as primary users (PU). In this case, the spectrum holes can be utilized by unlicensed users, also known as secondary users (SU) without introducing any interference to the PU [1]. This is popularly known as cognitive radio. The process of detecting spectrum holes is known as spectrum sensing in cognitive radio.

© ICST Institute for Computer Sciences, Social Informatics and Telecommunications Engineering 2021
Published by Springer Nature Switzerland AG 2021. All Rights Reserved
G. Caso et al. (Eds.): CrownCom 2020, LNICST 374, pp. 16–28, 2021.
https://doi.org/10.1007/978-3-030-73423-7_2

In the literature, various detection schemes for spectrum sensing with their performance in terms of receiver operating characteristic (ROC) is well documented. Increase in the sensing period can improve the detection performance but reduce the throughput [2]. Hence, there always exists a trade-off between the detection performance and the throughput [3]. The conventional spectrum sensing methods assume the static behaviour of PU during sensing period. It means that PU remains active or inactive throughout the whole sensing period [4–6]. In this scenario, it is a simple binary hypothesis testing problem, in which the assumed static model of PU is valid for slowly varying PU traffic such as in television broadcast [7]. However, when PU traffic varies fastly or PU is dynamic as in cellular communication or Wireless LAN [8,9], the performance of these schemes degrade.

The case of one transition of PU signal during spectrum sensing interval is considered in [4–9], while multiple transitions of PU signal were considered in [10,11]. In [12], two transitions of PU signal were considered during sensing period and dynamic programming was introduced to reduce the computation time. In [13], random arrival or random departure of PU is considered over the sensing period and it is assumed that the arrival and departure of the PU follows Poisson process. The effect of deep sensing is proposed in [14]. Two-state Markov chain is used in [15], where Markov model is used for the on-off status of the PU. Partially observed Markov decision process is discussed in [16], where occupancies of the vacant licensed band follows Markovian evolution. Cooperative spectrum sensing in dynamic PU environment is discussed in [17] with the aim of throughput maximization. The effect of dynamicity of the PU under both sensing period and transmission period is discussed in [18].

Furthermore, most of the schemes assume additive noise as Gaussian [12–18], which usually models thermal noise of the receiver. However, in the multiuser environment, dominant source of noise is interference such as multiple access interference (MAI). The MAI can be modelled by Gaussian mixture model (GMM), Middleton Class A model (MCA) or Laplacian noise model. The Laplacian model is used in [19] to model the distribution of MAI in time-hopped ultra-wideband (TH-UWB) communication system.

In this paper, we consider dynamic primary user (PU) with additive Laplacian noise [19,20]. We have used conventional energy detection along with absolute value cumulation detection (AVCD) [21] and improved AVCD (i-AVCD) [22] detection schemes. In AVCD, sum of the absolute values of the received samples over the sensing period is considered, while in case of i-AVCD, the sum of absolute value of each sample raised to a positive exponent P, $0 \leq P < 2$ is used. Thus, AVCD is a special case of i-AVCD with $P = 1$. Further, we have modelled the random arrival or the random departure of the PU using Poisson distribution. The arrival time or departure time of the PU is assumed to follow exponential distribution.

2   System Model

In general, when PU is static, H_o denotes the null hypothesis when PU is absent and H_1 denotes the alternate hypothesis when PU is present. Here, we assume that PU randomly departs under hypothesis H_o and randomly arrives under hypothesis H_1 during the sensing time. Received signals at the cognitive terminal under the random arrival and the random departure of the PU can be expressed as [13]

$$
\begin{aligned}
H_0 : y_m &= \begin{cases} s_m + w_m & , m = 1, \ldots, \xi_o \\ w_m & , m = \xi_o + 1, \xi_o + 2, \xi_o + 3 \ldots, N \end{cases} \\
H_1 : y_m &= \begin{cases} w_m & , m = 1, \ldots, \xi_1 \\ s_m + w_m & , m = \xi_1 + 1, \xi_1 + 2, \xi_1 + 3 \ldots, N \end{cases}
\end{aligned} \tag{1}
$$

where $m = 1, 2, \ldots, N$, N indicates the total number of samples present during the sensing period. $N = BT$, where B is the bandwidth of the bandpass filter used to filter PU signal and T is the interval at which filtered PU signal is sampled. s_m is the unknown PU signal, w_m indicates Laplacian noise with mean 0 and variance $2b^2$ where b is the scale parameter of Laplacian noise. Average SNR is defined as $\gamma = (1/N) \sum_{m=1}^{N} (s_m^2)/(2b^2)$. ξ_o and ξ_1 indicate the first level alter points of the PU under hypotheses H_o and H_1 respectively. Transition of the PU occurs between the samples ξ_o and $\xi_o + 1$ under H_o (when PU randomly departs) and between the samples ξ_1 and $\xi_1 + 1$ under H_1 (when PU randomly arrives). The PDF of Laplacian noise is expressed as

$$
f_{w_m}(x) = \frac{1}{2b} \exp\left(-\frac{|x|}{b}\right). \tag{2}
$$

3   Dynamic PU Modeling

In this section, we present P_D and P_F for different detection schemes such as energy detection (ED) and i-AVCD detection.

3.1   Energy Detection

The ED is the classical and the simplest spectrum sensing method in cognitive radio. From (1), the likelihood function under the null-hypothesis H_o can be expressed as

$$
f(\mathbf{y}|\mathbf{s}_{co}, H_o) = \frac{1}{(2b)^N} \exp\left\{ -\sum_{m=1}^{\xi_o} \frac{|y_m - s_m|^2}{b} - \sum_{m=\xi_o+1}^{N} \frac{|y_m|^2}{b} \right\}, \tag{3}
$$

where $\mathbf{y} = [y_1, y_2, \ldots, y_N]$ and $\mathbf{s}_{co} = [s_1, s_2, \ldots, s_{\xi_o}]$. Similarly from (1), the likelihood function under the hypothesis H_1 can be expressed as

$$f(\mathbf{y}|\mathbf{s}_{c1}, H_1) = \frac{1}{(2b)^N} \exp\left\{ -\sum_{m=1}^{\xi_1} \frac{|y_m|^2}{b} - \sum_{m=\xi_1+1}^{N} \frac{|y_m - s_m|^2}{b} \right\}, \quad (4)$$

where $\mathbf{y} = [y_1, y_2, \ldots, y_N]$ and $\mathbf{s}_{c1} = [s_{\xi_1+1}, s_{\xi_1+2}, \ldots, s_N]$. As the parameter s_m is unknown, CR is not aware of any information about PU signal. Hence, it needs to be removed from the likelihood function. This belongs to the composite hypothesis testing problem where some of the parameters in the hypotheses are unknown and can be estimated using various estimation techniques. More specifically, Generalized likelihood ratio test (GLRT) is used for spectrum sensing and maximum likelihood (ML) estimation is used to find out the estimate of s_m. This gives

$$\frac{f(\mathbf{y}|\hat{\mathbf{s}}_{c1}, H_1)}{f(\mathbf{y}|\hat{\mathbf{s}}_{co}, H_o)} = \frac{\frac{1}{(2b)^N} \exp\left\{ -\sum_{m=1}^{\xi_1} \frac{|y_m|^2}{b} \right\}}{\frac{1}{(2b)^N} \exp\left\{ -\sum_{m=\xi_o+1}^{N} \frac{|y_m|^2}{b} \right\}} \underset{H_o}{\overset{H_1}{\gtrless}} \lambda, \quad (5)$$

where $\hat{\mathbf{s}}_{co} = [\hat{s}_1, \hat{s}_2, \ldots, \hat{s}_{\xi_o}]$ and $\hat{\mathbf{s}}_{c1} = [\hat{s}_{\xi_1+1}, \hat{s}_{\xi_1+2}, \ldots, \hat{s}_N]$. Maximum likelihood (ML) estimate of s_m is $\hat{s}_m$ which is calculated and found to be y_m, i.e., $\hat{s}_m = y_m$. The values of ξ_o and ξ_1 are also random and unknown at this point. Hence, they should be averaged out over the likelihood ratio obtained in (5). Taking logarithm from both the sides in (5), the expression becomes

$$Z = \sum_{m=\xi_o+1}^{N} |y_m|^2 - \sum_{m=1}^{\xi_1} |y_m|^2 \underset{H_o}{\overset{H_1}{\gtrless}} \lambda', \quad (6)$$

where λ' is the detection threshold of decision statistic Z, that needs to be determined using Neyman-Pearson (NP) test. Detection Probability P_D and false alarm probability P_F can be expressed as

$$P_D = Pr\left\{ Z > \lambda'|H_1 \right\},$$

$$P_F = Pr\left\{ Z > \lambda'|H_o \right\}. \quad (7)$$

In this paper, the arrival rate and the departure rate of the PU is assumed to follow Poisson process. The arrival rate and departure rate of the PU is denoted as θ_A and θ_D, respectively. The probability of the PU for not arriving or departing during the sample interval T is given by $e^{-\theta_A T}$ and $e^{-\theta_D T}$, respectively. Hence, the probability of arrival or departure of the PU is given by by $1 - e^{-\theta_A T}$

and $1 - e^{-\theta_D T}$ [13], respectively. Thus, the probability of the random arrival or the random departure in the ξ_o^{th} or ξ_1^{th} sample is given by [13]

$$Pr\{\xi_o\} = \left\{ 1 - \exp\{\theta_D T\} \right\} \cdot \left\{ \exp\{-\theta_D T\} \right\}^{\xi_o},$$

$$Pr\{\xi_1\} = \left\{ 1 - \exp\{\theta_A T\} \right\} \cdot \left\{ \exp\{-\theta_A T\} \right\}^{\xi_1}, \tag{8}$$

where $Pr\{\xi_o\}$ and $Pr\{\xi_1\}$ denote the respective probability of random departure and random arrival, $\theta_A T$ and $\theta_D T$ denote the number of PU arrivals and departures, respectively. We consider three different decision statistics - ED, AVCD and i-AVCD. A case of static PU and two cases (arrival and departure) of dynamic PU are discussed in the following section using each of the three test statistics.

1) Static/Fixed PU: When PU is static or fixed at one place ED can be termed as conventional or classical ED. This denotes a case when $\xi_o = 0$ and $\xi_1 = 0$. When $\xi_o = 0$, PU is absent during the whole sensing period and when $\xi_1 = 0$, PU is present during the whole sensing period. This case complies with the case of low traffic scenario. Decision statistics under ED can be expressed as

$$C = \sum_{m=1}^{N} |y_m|^2 \underset{H_o}{\overset{H_1}{\gtrless}} \psi_c, \tag{9}$$

where ψ_c is the detection threshold of ED based decision statistic C obtained using NP test. For large values of N, central limit theorem (CLT) can be applied to approximate the probability density function (PDF) of C as Gaussian with mean m_o and variance σ_o^2, i.e.,

$$C \sim N(m_o, \sigma_o^2), \tag{10}$$

where $m_o = 2Nb^2$ and $\sigma_o = 2\sqrt{5N}b^2$ [22]. Using (10), ψ_c can be expressed as

$$\psi_c = Q^{-1}(P_F)\sigma_o + m_o, \tag{11}$$

where $Q(.)$ is the Q-function given by $Q(v) = \frac{1}{\sqrt{2\pi}} \int_v^{+\infty} \exp\left(-\frac{t^2}{2}\right) dt$.

2) Dynamic PU (Random Arrival): It is now known that when $\xi_o = 0$, PU is absent during the whole sensing period. However, it also signifies the initial phase of the potential transmission of the PU. Using (6) and (8), averaged likelihood ratio (ALR) can be obtained as [13]

$$A = \sum_{\xi_1=0}^{N-1} \left\{ 1 - \exp\{\theta_A T\} \right\} \left\{ \exp\{-\theta_A T\} \right\}^{\xi_1} \left[\sum_{m=1}^{N} |y_m|^2 - \sum_{m=1}^{\xi_1} |y_m|^2 \right] \underset{H_o}{\overset{H_1}{\gtrless}} \psi_a$$

$$= \sum_{m=1}^{N} \left\{ 1 - \exp\{-\theta_A T m\} \right\} |y_m|^2 \underset{H_o}{\overset{H_1}{\gtrless}} \psi_a, \tag{12}$$

where ψ_a is the detection threshold of ED based decision statistic A when PU arrives in random fashion. Applying CLT, A tends to be Gaussian with mean μ_A and variance σ_A^2. Thus, ψ_A can be expressed as

$$\psi_a = Q^{-1}(P_F)\sigma_A + \mu_A, \tag{13}$$

where μ_A and σ_A^2 denote the mean and the variance, respectively of the decision statistic A obtained in (12) under hypothesis H_o. The expression of μ_A and σ_A^2 can be derived and expressed as

$$\mu_A = N\mu_o - \mu_o \left[\frac{\exp(-\theta_A T)\{1 - \exp(-\theta_A TN)\}}{1 - \exp(-\theta_A T)} \right],$$

$$\sigma_A^2 = N\sigma_o^2 - \sigma_o^2 \left[\frac{\exp(-2(\theta_A T + 1))\{1 - \exp(-(N-1))\}}{1 - \exp(-1)} + \exp(-2\theta_A T) \right], \tag{14}$$

where μ_o and σ_o^2 are the mean and variance of the decision statistic C based on conventional/classical ED in Laplacian noise. They can be obtained from (11) with $\mu_o = 2Nb^2$ and $\sigma_o^2 = 20Nb^4$.

3) Dynamic PU (Random Departure): When $\xi_1 = 0$, PU is present during the whole sensing period. However, it also signifies the final or last phase of the active transmission of the PU. Using (6) and (8), ALR can be obtained as [13]

$$D = \sum_{\xi_o=0}^{N-1} \{1 - \exp\{\theta_D T\}\}\{\exp\{-\theta_D T\}\}^{\xi_o} \sum_{m=\xi_o+1}^{N} |y_m|^2$$

$$= \sum_{m=1}^{N} \{1 - \exp\{-\theta_D Tm\}\}|y_m|^2 \underset{H_o}{\overset{H_1}{\gtrless}} \psi_d, \tag{15}$$

where ψ_d is the detection threshold of decision statistic D based on ED when PU departs in random fashion. Applying CLT, D tends to be Gaussian with mean μ_D and variance σ_D^2. Thus, ψ_D can be expressed as and it can be expressed as

$$\psi_d = Q^{-1}(P_F)\sigma_D + \mu_D, \tag{16}$$

where μ_D and σ_D^2 denote the mean and the variance, respectively, of the decision statistic D obtained in (15) under hypothesis H_o. Expression of μ_D and σ_D^2 is derived and expressed as

$$\mu_D = N\mu_o - \mu_o \left[\frac{\exp(-\theta_D T)\{1 - \exp(-\theta_D TN)\}}{1 - \exp(-\theta_D T)} \right],$$

$$\sigma_D^2 = N\sigma_o^2 - \sigma_o^2 \left[\frac{\exp(-2(\theta_D T + 1))\{1 - \exp(-(N-1))\}}{1 - \exp(-1)} + \exp(-2\theta_D T) \right]. \tag{17}$$

3.2   AVCD and i-AVCD

AVCD and i-AVCD are the two important test-statistics used actively in Laplacian noise. As discussed before, AVCD is a special case of i-AVCD with exponent $P = 1$. Hence, in this section we particularly describe i-AVCD. Corresponding parameters of AVCD can be obtained by putting the value of $P = 1$ in the expressions obtained for i-AVCD. Removing the squaring operation and substituting it with P in (3)–(5), decision statistics can be expressed as

$$Z = \sum_{m=\xi_o+1}^{N} |y_m|^P - \sum_{m=1}^{\xi_1} |y_m|^P \underset{H_o}{\overset{H_1}{\gtrless}} \lambda', \tag{18}$$

where λ' is the detection threshold of the test-statistic Z obtained in (18). Three different cases under dynamic modeling of i-AVCD based spectrum sensing are discussed in the following section.

1) Static PU: When PU is static, i-AVCD can be termed as conventional or classical i-AVCD. This denotes the case when $\xi_o = 0$, $\xi_1 = 0$. The decision statistic under this case can be expressed as

$$C = \sum_{m=1}^{N} |y_m|^P \underset{H_o}{\overset{H_1}{\gtrless}} \psi_c, \tag{19}$$

where ψ_c is the detection threshold of i-AVCD based decision statistic C obtained in (19) using NP test and it can be expressed as

$$\psi_c = Q^{-1}(P_F)\sigma_o + \mu_o, \tag{20}$$

where μ_o and σ_o^2 denote the mean and the variance, respectively of the decision statistic C obtained in (19) expressed as [22]

$$\mu_o = b^P \Gamma(P+1),$$
$$\sigma_o^2 = b^{2P} \Gamma(2P+1) - \mu_o^2, \tag{21}$$

where $\Gamma(v) = \int_0^{+\infty} e^{-t} t^{v-1} dt$.

2) Dynamic PU (Random Arrival): Similar to the ED, in this case, the ALR can be obtained as

$$A = \sum_{\xi_1=0}^{N-1} \{1 - \exp\{\theta_A T\}\}\{\exp\{-\theta_A T\}\}^{\xi_1} \left[\sum_{m=1}^{N} |y_m|^P - \sum_{m=1}^{\xi_1} |y_m|^P \right] \underset{H_o}{\overset{H_1}{\gtrless}} \psi_a,$$
$$= \sum_{m=1}^{N} \{1 - \exp\{-\theta_A T m\}\} |y_m|^P \underset{H_o}{\overset{H_1}{\gtrless}} \psi_a, \tag{22}$$

where ψ_a is the detection threshold of i-AVCD based decision statistic A, when PU arrives randomly which can be expressed as

$$\psi_a = Q^{-1}(P_F)\sigma_A + \mu_A, \tag{23}$$

where μ_A and σ_A^2 denote the mean and the variance respectively, of the decision statistic obtained in (22) under hypothesis H_o. The expression of μ_A and σ_A^2 can be derived and expressed as

$$\mu_A = N\mu_o - \mu_o \left[\frac{\exp(-\theta_A T)\{1 - \exp(-\theta_A TN)\}}{1 - \exp(-\theta_A T)} \right],$$

$$\sigma_A^2 = N\sigma_o^2 - \sigma_o^2 \left[\frac{\exp(-2(\theta_A T + 1))\{1 - \exp(-(N-1))\}}{1 - \exp(-1)} + \exp(-2\theta_A T) \right], \tag{24}$$

3) Dynamic PU (Random Departure): In this case, the ALR can be obtained as [13]

$$D = \sum_{\xi_o=0}^{N-1} \{1 - \exp\{\theta_D T\}\}\{\exp\{-\theta_D T\}\}^{\xi_o} \sum_{m=\xi_o+1}^{N} |y_m|^P \underset{H_o}{\overset{H_1}{\gtrless}} \psi_d$$

$$= \sum_{m=1}^{N} \{1 - \exp\{-\theta_D Tm\}\}|y_m|^P \underset{H_o}{\overset{H_1}{\gtrless}} \psi_d, \tag{25}$$

where ϕ_d is the detection threshold of i-AVCD based decision statistic D, obtained in (25), when PU departs randomly. It can be expressed as

$$\psi_d = Q^{-1}(P_F)\sigma_D + \mu_D, \tag{26}$$

where μ_D and σ_D^2 denote the mean and the variance respectively of the decision statistic, obtained in (25) under hypothesis H_o. The expressions of μ_D and σ_D^2 can be derived and expressed as

$$\mu_D = N\mu_o - \mu_o \left[\frac{\exp(-\theta_D T)\{1 - \exp(-\theta_D TN)\}}{1 - \exp(-\theta_D T)} \right],$$

$$\sigma_D^2 = N\sigma_o^2 - \sigma_o^2 \left[\frac{\exp(-2(\theta_D T + 1))\{1 - \exp(-(N-1))\}}{1 - \exp(-1)} + \exp(-2\theta_D T) \right]. \tag{27}$$

The mean and variance, obtained under dynamic scenario, are dependent on the parameters T and N. The random arrival and random departure are dependent on the respective arrival rate θ_A and departure rate θ_D, total number of sensing samples N and sample interval T.

4   Results

In this section, performance of the proposed spectrum sensing in dynamic PU environment with additive Laplacian noise is presented in terms of receiver operating characteristic (ROC) and detection probability versus average signal-to-noise ratio (SNR) γ using Monte Carlo simulations. The values of ξ_o and ξ_1 are taken to be 10 and 15 respectively. For static PU environment, $\xi_o = 0$ and $\xi_1 = 0$, whereas for dynamic PU environment both are less than N.

Figure 1 shows the ROC for ED, AVCD and i-AVCD (for $P = 0.8$) under both static and dynamic scenarios of the PU at $\gamma = -5$ dB, $\theta_A T = 1$, $\theta_D T = 1$ and $N = 50$. It can be seen that the performance in the dynamic scenario is better than the performance in the static scenario in all the three schemes. We have further observed that the performance with i-AVCD improves, when P reduces in the range $0 \leq P < 2$.

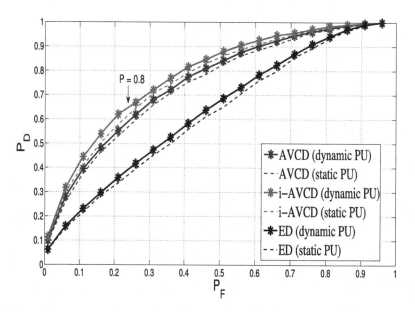

Fig. 1. Comparison of ROC plot for different test statistics with $N = 50$, $\gamma = -5$ dB, $\theta_A T = 1$ and $\theta_D T = 1$.

Figure 2 presents the detection probability (P_D) versus average SNR γ for all the three schemes: ED, AVCD and i-AVCD for $P_F = 0.1$, $\theta_A T = 1$, $\theta_D T = 1$ and $N = 50$. Here, also i-AVCD outperforms other two schemes.

Figure 3 shows ROC for AVCD scheme under different dynamic scenario with different values of $\theta_A T$ and $\theta_D T$ at SNR $\gamma = -2$ dB and $N = 50$. For $\theta_A T$ and $\theta_D T$ as 0.05, 0.1, 1, 10 and 20. We have also presented performance with static scenario also. It can be seen that the performance will improve as value of $\theta_A T$

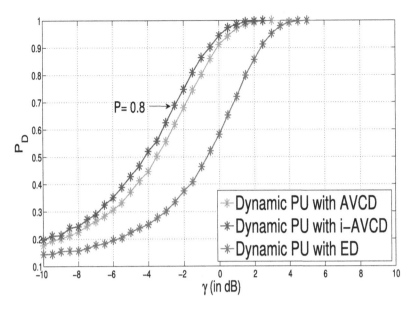

Fig. 2. P_D vs average SNR comparison with $N = 50$ and $P_F = 0.1$, $\theta_A T = 1$ and $\theta_D T = 1$ for randomly arriving or departing PU under different test statistic.

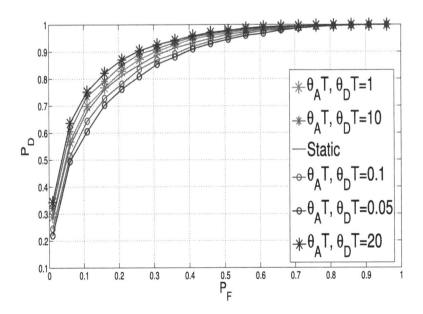

Fig. 3. P_D vs P_F plot at $N = 50$ and $\gamma = -2$ dB for randomly arriving or departing PU with AVCD based test statistic.

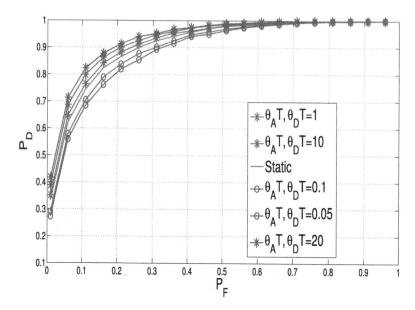

Fig. 4. P_D vs P_F plot at $N = 50$, $P = 0.8$ and $\gamma = -2$ dB for randomly arriving or departing PU with i-AVCD based test statistic.

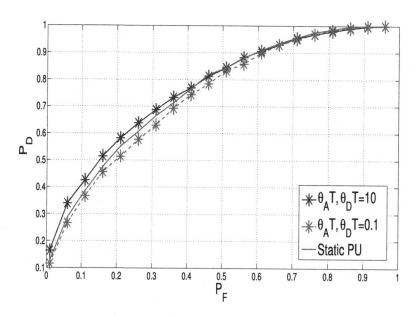

Fig. 5. P_D vs P_F plot at $N = 50$ and $\gamma = -2$ dB for randomly arriving or departing PU with ED based test statistic.

and $\theta_D T$ increases from 0.05 to 20. However, the performance for $\theta_A T$ and $\theta_D T$ of below 1 is worse than the performance with static scenario.

Similar to the above figure, in Fig. 4, we represent detection performance of i-AVCD based test-statistic for different values of $\theta_A T$ and $\theta_D T$ from 0.05 to 20 with $P = 0.8$, SNR $\gamma = -2\,\mathrm{dB}$ and $N = 50$. We have observed the similar effect. Compared to the static case, the performance in dynamic case with $\theta_A T$ and $\theta_D T$ of 1, 10 and 20 is better. However, the performance with $\theta_A T$ and $\theta_D T$ of 0.05 and 0.1 is worse than the static case.

Similar the above-mentioned two figures, Fig. 5 shows the ROC performance with ED under both dynamic and static PU condition. We have again observed the similar trend.

5   Conclusion

In this paper, we considered spectrum sensing schemes such as energy detection, absolute value cumulation detection (AVCD) and improved AVCD (i-AVCD) in the additive Laplacian noise environment. Further, we considered the dynamic behaviour of primary user by assuming its random arrival and/or departure (in terms of $\theta_A T$ and/or $\theta_D T$) in the sensing interval. We present the performance using simulations in terms of receiver operating characteristics and detection probability versus average SNR. We conclude that the performance with dynamic scenario with $\theta_A T$ and/or $\theta_D T$ beyond 1 is better than the static scenario.

References

1. Awin, F.A., Alginahi, Y.M., Abdel-Raheem, E., Tepe, K.: Technical issues on cognitive radio-based internet of things systems: a survey. IEEE Access **7**, 97887–97908 (2019). https://doi.org/10.1109/ACCESS.2019.2929915
2. Ali, A., Hamouda, W.: Advances on spectrum sensing for cognitive radio networks: theory and applications. IEEE Commun. Surv. Tutor. **19**(2), 1277–1304 (2017). https://doi.org/10.1109/COMST.2016.2631080
3. Yang, L., Fang, J., Duan, H., Li, H.: Fast compressed power spectrum estimation: toward a practical solution for wideband spectrum sensing. IEEE Trans. Wirel. Commun. **19**(1), 520–532 (2020). https://doi.org/10.1109/TWC.2019.2946805
4. Liu, M., Zhao, N., Li, J., Leung, V.C.M.: Spectrum sensing based on maximum generalized correntropy under symmetric alpha stable noise. IEEE Trans. Veh. Technol. **68**(10), 10262–10266 (2019). https://doi.org/10.1109/TVT.2019.2931949
5. Zou, Y., Yao, Y., Zheng, B.: Outage probability analysis of cognitive transmissions: impact of spectrum sensing overhead. IEEE Trans. Wirel. Commun. **9**(8), 2676–2688 (2010). https://doi.org/10.1109/TWC.2010.061710.100108
6. Chen, Y.: Improved energy detector for random signals in Gaussian noise. IEEE Trans. Wirel. Commun. **9**(2), 558–563 (2010). https://doi.org/10.1109/TWC.2010.5403535
7. Cavalcanti, D., Ghosh, M.: Cognitive radio networks: enabling new wireless broadband opportunities. In: 2008 3rd International Conference on Cognitive Radio Oriented Wireless Networks and Communications, pp. 1–6, May 2008. https://doi.org/10.1109/CROWNCOM.2008.4562540

8. Csurgai Horvath, L., Bito, J.: Primary and secondary user activity models for cognitive wireless network. In: Proceedings of the 11th International Conference on Telecommunications, pp. 301–306, June 2011

9. Geirhofer, S., Tong, L., Sadler, B.M.: Cognitive medium access: constraining interference based on experimental models. IEEE J. Sel. Areas Commun. **26**(1), 95–105 (2008). https://doi.org/10.1109/JSAC.2008.080109

10. Pradhan, H., Kalamkar, S.S., Banerjee, A.: Sensing-throughput tradeoff in cognitive radio with random arrivals and departures of multiple primary users. IEEE Commun. Lett. **19**(3), 415–418 (2015). https://doi.org/10.1109/LCOMM.2015. 2393305

11. Tang, L., Chen, Y., Hines, E.L., Alouini, M.: Performance analysis of spectrum sensing with multiple status changes in primary user traffic. IEEE Commun. Lett. **16**(6), 874–877 (2012). https://doi.org/10.1109/LCOMM.2012.041112.120507

12. Düzenli, T., Akay, O.: A new spectrum sensing strategy for dynamic primary users in cognitive radio. IEEE Commun. Lett. **20**(4), 752–755 (2016). https://doi.org/10.1109/LCOMM.2016.2527640

13. Beaulieu, N.C., Chen, Y.: Improved energy detectors for cognitive radios with randomly arriving or departing primary users. IEEE Signal Process. Lett. **17**(10), 867–870 (2010). https://doi.org/10.1109/LSP.2010.2064768

14. Li, B., Hou, J., Li, X., Nan, Y., Nallanathan, A., Zhao, C.: Deep sensing for space-time doubly selective channels: when a primary user is mobile and the channel is flat Rayleigh fading. IEEE Trans. Signal Process. **64**(13), 3362–3375 (2016). https://doi.org/10.1109/TSP.2016.2537276

15. Unnikrishnan, J., Veeravalli, V.V.: Algorithms for dynamic spectrum access with learning for cognitive radio. IEEE Trans. Signal Process. **58**(2), 750–760 (2010). https://doi.org/10.1109/TSP.2009.202

16. MacDonald, S., Popescu, D.C., Popescu, O.: Analyzing the performance of spectrum sensing in cognitive radio systems with dynamic PU activity. IEEE Commun. Lett. **21**(9), 2037–2040 (2017). https://doi.org/10.1109/LCOMM.2017.2705126

17. Yilmaz, Y., Guo, Z., Wang, X.: Sequential joint spectrum sensing and channel estimation for dynamic spectrum access. IEEE J. Sel. Areas Commun. **32**(11), 2000–2012 (2014). https://doi.org/10.1109/JSAC.2014.141105

18. Chang, K., Senadji, B.: Spectrum sensing optimisation for dynamic primary user signal. IEEE Trans. Commun. **60**(12), 3632–3640 (2012). https://doi.org/10.1109/TCOMM.2012.091712.110856

19. Win, M.Z., Scholtz, R.A.: Ultra-wide bandwidth time-hopping spread-spectrum impulse radio for wireless multiple-access communications. IEEE Trans. Commun. **48**(4), 679–689 (2000). https://doi.org/10.1109/26.843135

20. Hu, B., Beaulieu, N.C.: On characterizing multiple access interference in TH-UWB systems with impulsive noise models. In: 2008 IEEE Radio and Wireless Symposium, pp. 879–882, January 2008. https://doi.org/10.1109/RWS.2008.4463633

21. Ye, Y., Li, Y., Lu, G., Zhou, F., Zhang, H.: Performance of spectrum sensing based on absolute value cumulation in Laplacian noise. In: 2017 IEEE 86th Vehicular Technology Conference (VTC-Fall), pp. 1–5, September 2017. https://doi.org/10. 1109/VTCFall.2017.8287978

22. Ye, Y., Li, Y., Lu, G., Zhou, F.: Improved energy detection with Laplacian noise in cognitive radio. IEEE Syst. J. **13**(1), 18–29 (2019). https://doi.org/10.1109/JSYST.2017.2759222

Blind Source Separation for Wireless Networks: A Tool for Topology Sensing
(Invited Paper)

Enrico Testi[✉], Elia Favarelli, and Andrea Giorgetti

Alma Mater Studiorum – University of Bologna, Via dell'Università 50, Cesena, Italy
{enrico.testi4,elia.favarelli2,andrea.giorgetti}@unibo.it

Abstract. In this work, a tool for topology sensing of a non-collaborative wireless network using power profiles captured by radio-frequency (RF) sensors is proposed. Assuming that the features of the network (i.e., the number of nodes, medium access control (MAC) and routing protocols) are unknown and that the sensors observe signal mixtures because of the wireless medium, blind source separation (BSS) is used to separate the traffic profiles. Successively, the topology of the network is inferred by detecting causal relationships between the separated streams. According to the numerical results, the proposed tool senses the topology with promising accuracy when operating in mild shadowing conditions, even with a relatively low number of radio-frequency (RF) sensors.

Keywords: Blind source separation · Topology sensing · Wireless networks · Cognitive radio

1 Introduction

Thanks to the impressive advances in research in communication technologies, networks' importance is strongly growing within our society. A network of sensors deployed to collect environmental data, a community of social network users, or a tactical network suited for the exchange of data between soldiers are only a small part of the vastness of common networks.

In this scenario, to predict the traffic flow, to estimate the network connectivity, detect communities, infer the communications between nodes, help optimization and orchestration are only a few examples of tasks that can benefit the topology sensing. Focusing on wireless networks, while cognitive radios (CRs) are currently exploiting spectrum sensing, a detailed knowledge on how a network uses the radio-frequency (RF) medium can be obtained by the structure of the network, and it might help the development of more performing spectrum sharing algorithms [1,4,20,22].

In many of the scenarios mentioned above, it seems mandatory that the topology of the network is inferred from external, without being a part of it

© ICST Institute for Computer Sciences, Social Informatics and Telecommunications Engineering 2021
Published by Springer Nature Switzerland AG 2021. All Rights Reserved
G. Caso et al. (Eds.): CrownCom 2020, LNICST 374, pp. 29–42, 2021.
https://doi.org/10.1007/978-3-030-73423-7_3

or increasing the network overhead. For this reason, the possibility of reconstructing the topology of a network from observed signals at some nodes with almost zero prior knowledge is being investigated nowadays [10,28,33]. Assuming that the main features of the network, such as the number of nodes, their accurate position, and the traffic types are unknown, our objective is to infer its topology by observing it from external. Firstly, an estimate of the number of transmitting nodes (sources) is provided, then the transmitted power profiles as if they were measured at the sensors are separated using blind source separation (BSS). Finally, we compare well-known topology inference tools based on Granger causality (GC) and transfer entropy (TE) accounting for noise and propagation impairments.

1.1 Existing Works

In literature, numerous methodologies for network topology inference have been proposed and tested. Some of them require the packet's content to be accessible, which might not always be granted, and increment the total network overhead [10,28]. Others require access to information at endpoints, falling into the network tomography category [31,33].

The topology sensing problem can be tackled using statistical tests built on temporal models that relate different time series, as theorized by Pearl [16] and Granger [5]. In particular, the auto-regressive (AR) models based test proposed by Granger in [5], became the reference for numerous causal analysis and inference methodologies proposed in the last decades. As an alternative, a technique to infer causal relationships within networks based on a variant of the classic GC, named asymmetric Granger causality (AGC), is exploited in [11]. A worth-mentioning technique to model causal relationships is represented by the Hawkes point processes [30]. Another state-of-the-art methodology for causality detection is TE, which exploits the amount of directed transfer of information between two time series [18]. In [19], a TE-based topology sensing method is proposed and compared to the GC test. The presented topology inference methods require the temporal characterization of the signals transmitted by each node of the network. For this reason, the packet streams have to be reconstructed as if they were measured at the nodes. Due to the fact that the RF sensors acquire a mixture of the signals generated by the nodes because of the wireless medium, an unmixing operation to extract the transmitted power profiles is necessary [9,32].

Various methods for unmixing signals, e.g., matrix factorization [3] and tensor decomposition [2,8], have been proposed in literature. In this work, we combine the principal component analysis (PCA) and Fast-independent component analysis (ICA) methodologies [6]. We then propose a specific solution to associate the reconstructed sequences and compare some of the topology inference algorithms mentioned above. In Sect. 5, we inspect how shadowing impacts the reconstruction of the packet profiles, and, consequently, the reliability of the sensed topology [23,24].

1.2 Application Scenarios

The designed tool senses the topology of a wireless network, assuming that it is non-collaborative. This could be necessary, i.e., because it is private, encrypted, and not accessible, or it is a contender in the use of shared spectrum. This is the reason why the BSS and the topology sensing have to be carried out from the outside. To motivate this assumption, two generic examples of application are briefly described in the following:

– Spectrum optimization and reuse: let us consider a wireless network aiming to detect if another network is competing for the use of the same portion of the radio spectrum (also not legally). The nodes repeatedly sense the RF spectrum [7,12] to collect power samples and forward them to the fusion center (FC). It is in charge of the detection of the competing wireless network, performing BSS and topology inference. Once the adversarial topology is sensed, the wireless network is capable of optimizing its communications with respect to the structure and the spectrum usage of the competing network. Of course, the time necessary to probe the radio spectrum worsens the throughput of the network while optimizing the dynamic spectrum allocation.
– Tactical network: in this scenario, a network of RF sensors is deployed within an adversarial network's area to obtain information about it. The sensors acquire the radio spectrum samples and forward them to the designed FC. It performs BSS and topology inference. If the specifications of the sensor network (i.e., the energy consumption) allow it, the topology can be inferred in real-time.

Throughout the paper, capital boldface letters denote matrices, lowercase bold letters denote vectors, $(\cdot)^{\mathrm{T}}$ is the transpose operator, $\odot$ is the element-wise product, $|\cdot|$ is the modulus operator, and $||\cdot||_p$ is the l_p-norm. With $\mathbf{v}_{h,:}$ and $\mathbf{v}_{:,g}$ we represent the hth row, and the gth column of the matrix $\mathbf{V}$ respectively, and with $\mathbf{v}_{h,g:k}$ we select the elements between the gth and the kth entry of the hth row of $\mathbf{V}$. $\mathcal{N}(\mu, \sigma^2)$ represents a real normal distribution with mean μ and variance σ^2, $\mathbb{E}\{\cdot\}$ returns the expected value, and $\langle\cdot\rangle$ returns the sample mean.

The rest of this work is structured as follows. The system model is introduced in Sect. 2. In Sect. 3, the blind source separation problem and the proposed solutions are described in detail. In Sect. 4, two of the state-of-the-art techniques for topology inference mentioned in Sect. 1.1 are presented. Numerical results are shown in Sect. 5 and conclusions are deduced in Sect. 6.

2 System Model and Problem Formulation

Let us consider a network of M RF sensors and an adversarial (non-collaborative) wireless network of N nodes deployed within a square area of side L_{a}.[1] Let us assume that the network's technical specifications (i.e., number of nodes, routing

---

[1] The BSS requires a coarse estimate of the network nodes position, therefore it can be performed by the sensor network before the tasks summarized in Fig. 1 [14,15].

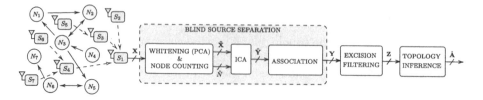

Fig. 1. Block scheme of the proposed tool: sensor network, BSS and topology inference.

protocols) are unknown. Usually, the network's topology can be seen as a directed graph and represented as its associated adjacency matrix $\mathbf{A} \in \{0,1\}^{N \times N}$ where

$$a_{h,k} = \begin{cases} 1 & \text{if the data flow from node } h \text{ to } k \\ 0 & \text{otherwise.} \end{cases}$$

The objective is to estimate the adjacency matrix, $\widehat{\mathbf{A}}$, of the wireless network from the power profiles captured by the RF sensors within an observation window of duration T_{ob}. The sensors cannot interact with the observed network. Therefore, to perform the data processing summarized in Fig. 1, the demodulation of the received signals is not required, so that a simple energy detector (ED) is needed at the receiver [12,26].

2.1 Data Acquisition and Channel Model

The streams of packets transmitted by each node of the network are time series arranged as rows of the matrix $\mathbf{P} \in \mathbb{R}^{N \times K}$ where K is the number of power samples collected. The channel is represented by the mixing matrix $\mathbf{H} \in \mathbb{R}^{M \times N}$, whose entries are the channel gains $h_{m,n}$ between sensor m and node n.

The effect of shadowing between the nodes and the sensors is modeled by a log-normal distributed random matrix $\mathbf{S}$ whose elements are

$$s_{m,n} = \exp(\sigma_S G_{m,n}) \quad m = 1, \ldots, M, \ n = 1, \ldots, N$$

where $G_{m,n}$ are independent, identically distributed (i.i.d.) zero mean Gaussian random variables (r.v.s) with unit variance and shadowing parameter σ_S.[2] Moreover, assuming conventional energy detection, the thermal noise at the output of the integrator can be modeled as a constant ν added to each received sample.[3]

---

[2] Usually, the shadowing parameter is expressed as the standard deviation of the received power in deciBel, i.e., $\sigma_S(\text{dB}) = \frac{10}{\ln 10}\sigma_S$.

[3] The noise term in the ED is a central chi-squared r.v. with a number of degrees of freedom, N_{DOF}, proportional to the time-bandwidth product. When N_{DOF} is large, the noise term can be considered constant. As detailed in Sect. 5 we consider a system with bandwidth $W = 20\,\text{MHz}$ (i.e., WiFi channel) and an integration time $T_i = 10\,\mu\text{s}$, hence $N_{DOF} = 2WT_i = 400$.

Therefore, the matrix of the received power profile $\mathbf{X} \in \mathbb{R}^{M \times K}$ can be written as

$$\mathbf{X} = (\mathbf{S} \odot \mathbf{H})\mathbf{P} + \nu \mathbf{1}_{M,K} \tag{1}$$

where $\mathbf{1}_{M,K}$ is an $M \times K$ matrix of all ones. Note that the matrix $\mathbf{P}$ of the transmitted traffic profiles of the wireless network is somehow implicitly contained in $\mathbf{X}$. Nevertheless, the extraction of the desired packet streams is challenging because of physical layer impairments (i.e., propagation and noise), packet collisions, and multiple access interference.

3   Blind Source Separation

BSS is a statistical tool which purpose is to unmix the source signals (i.e., $\mathbf{P}$) from the mixture of observations (i.e., $\mathbf{X}$), when the mixing matrix (i.e., $\mathbf{H}$) is unknown [32]. Here we assume that the problem is overdetermined, that is $M \geq N$ (more sensors than sources).

3.1   Whitening and Estimation of the Number of Sources

For signal unmixing to be effective, data have to be manipulated so that there are N whitened mixtures [32]. Since we assumed that the number of nodes N is unknown, it has to be estimated.

Firstly, the row-wise mean is subtracted from $\mathbf{X}$, centering the mixtures. Then, a linear transformation named PCA is applied to the signals to whiten their components. Given the sample covariance matrix of the observations $\boldsymbol{\Sigma} = \frac{1}{K}\mathbf{X}\mathbf{X}^{\mathrm{T}}$, the eigenvalue decomposition $\boldsymbol{\Sigma} = \mathbf{U}\boldsymbol{\Lambda}\mathbf{U}^{\mathrm{T}}$ is performed. $\mathbf{U}$ is the eigenvectors matrix, and $\boldsymbol{\Lambda}$ is the diagonal matrix of the eigenvalues, Λ_i, with $i = 1, \ldots, M$, sorted in descending order. Then, the whitening matrix can be written as $\mathbf{Q} = \boldsymbol{\Lambda}^{-\frac{1}{2}}\mathbf{U}^{\mathrm{T}}$. The number of sources $\hat{N}$ is estimated using the minimum description length (MDL) criteria proposed in [13,27], such that

$$\hat{N} = \underset{n \in \{1,..,M\}}{\arg\min} \{\mathsf{MDL}(n)\} \tag{2}$$

where n is the unknown model order. Once $\hat{N}$ is estimated, the mixture is projected onto a new subspace which dimensionality is reduced from M to $\hat{N}$. This is accomplished by a projection matrix $\tilde{\mathbf{Q}}$ obtained from the first $\hat{N}$ rows of $\mathbf{Q}$, so that the whitened mixture can be written as $\tilde{\mathbf{X}} = \tilde{\mathbf{Q}}\mathbf{X}$.

3.2   Independent Component Analysis

ICA is a well-known method which goal is to find a linear representation of non-gaussian data so that the components are statistically independent. Here we apply ICA to unmix the transmitted packet streams (implicitly contained in $\mathbf{P}$). ICA allows us to estimate the unmixing matrix $\mathbf{W} \in \mathbb{R}^{\hat{N} \times \hat{N}}$ such that

$$\tilde{\mathbf{Y}} = \mathbf{W}^{\mathrm{T}}\tilde{\mathbf{X}} \tag{3}$$

Algorithm 1: Unmixed signals association

In : Separated components $\tilde{\mathbf{Y}}$, $\hat{N}$, $\mathbf{D}$
Out: Aligned reconstructed power profiles $\mathbf{Y}$

1 **for** n *from* 1 *to* $\hat{N}$ **do**
2 $\quad$ $h \leftarrow \arg\min_m \{d_{m,n}\}$
3 $\quad$ **for** k *from* 1 *to* length *of* $\tilde{\mathbf{y}}_{:,1}$ **do**
4 $\quad\quad$ $peaks_k \leftarrow \max\{\mathrm{corr}(\tilde{\mathbf{y}}_{k,:}; \mathbf{x}_{h,:})\}^{\dagger}$
5 $\quad$ **end**
6 $\quad$ $p \leftarrow \arg\max_k \{\mathbf{peaks}\}$
7 $\quad$ $\mathbf{Y}_{n,:} \leftarrow \tilde{\mathbf{y}}_{p,:}$
8 $\quad$ $\tilde{\mathbf{Y}} \leftarrow \tilde{\mathbf{Y}}/\tilde{\mathbf{y}}_{p,:}{}^{\ddagger}$
9 **end**

---

† $\mathrm{corr}(\mathbf{c}; \mathbf{d})$ returns the cross-correlation between $\mathbf{c}$ and $\mathbf{d}$. ‡ $\tilde{\mathbf{Y}}/\tilde{\mathbf{y}}_{p,:}$ removes the pth row from $\tilde{\mathbf{Y}}$.

where $\tilde{\mathbf{Y}} \in \mathbb{R}^{\hat{N}\times K}$ is the matrix of the separated signals. Due to the particularity of the application here we propose the well-known Fast-ICA iterative algorithm, with kurtosis as a non-linear function, and decorrelation based on the Gram-Schmidt scheme [6].

Unfortunately, ICA presents an ambiguity in the order of the recovered signals; thus $\mathbf{P}$ could be obtained from $\tilde{\mathbf{Y}}$ permuting its rows. An ad-hoc solution to this problem is proposed in the next section.

3.3   Unmixed Signals Association

We propose an iterative method to match the reconstructed signals to the nodes of the wireless network. For convenience, the matrix $\mathbf{D} \in \mathbb{R}^{M\times\hat{N}}$, whose entries $d_{m,n}$ are the two-dimensional distances between sensor m and node n, is defined. The proposed algorithm is the following:

– Firstly, a node n is selected from the nodes of the wireless network and its nearest sensor m is found.
– Then, the signal acquired by sensor m is correlated with all the reconstructed signals (rows of $\tilde{\mathbf{Y}}$) separately.
– The row $\tilde{\mathbf{y}}_{p,:}$ showing the highest positive correlation peak matches with node n and is copied into the nth row of $\mathbf{Y}$, accordingly.
– Finally, the matched node is removed from the list of all the possible nodes, the pth sequence is deleted from $\tilde{\mathbf{Y}}$ and the algorithm is iterated from the beginning.

The algorithm is further detailed in Algorithm 1. Its complexity results $\mathcal{O}(\hat{N}\log\hat{N})$, making it also applicable to networks with a large number of nodes.

3.4   Excision Filter

The output $\mathbf{Y}$ of the signal association has residual crosstalk due to the presence of noise and shadowing, and it has to be removed, e.g., by an excision filter.

The signals in $\mathbf{Y}$ have been processed to obtain sequences of 0s and 1s arranged in a matrix $\mathbf{Z}$. In particular, the element $z_{n,k}$ contains 1 if node n is sending a packet at the time sample k and 0 otherwise. To do so, we use the received samples to detect the event "packet sent" by the conventional binary hypothesis test. The threshold λ_n is set as a fraction $q \in [0,1]$ of the maximum of $\mathbf{y}_{n,:}$, i.e.,

$$\lambda_n = q \cdot \max_k \{y_{n,k}\}, \quad n = 1, \ldots, N. \tag{4}$$

4   Topology Inference Algorithms

In this section, the topology sensing methodologies, namely GC and TE, adopted in this work are briefly presented.

4.1   Granger Causality

GC method relies on linear L-order AR models. Considering two rows of $\mathbf{Z}$ corresponding to the time series of the transmitted power streams of nodes i and j, respectively, $\mathbf{z}_i$ and $\mathbf{z}_j$, two AR models can be identified

$$\mathcal{H}_1 : z_{j,k} = \sum_{l=1}^{L} \beta_l z_{j,k-l} + \sum_{l=1}^{L} \gamma_l z_{i,k-l} + \varepsilon_k \tag{5}$$

$$\mathcal{H}_0 : z_{j,k} = \sum_{l=1}^{L} \delta_l z_{j,k-l} + \omega_k \tag{6}$$

where $\{\beta_l\}_{l=1}^{L}$, $\{\gamma_l\}_{l=1}^{L}$, and $\{\delta_l\}_{l=1}^{L}$ are the regression coefficients, and ε_k, ω_k, are samples of independent additive white Gaussian noise (AWGN). The model (5) is formulated according to hypothesis $\mathcal{H}_1$, considering the possibility of a causal relationship between the two traffic streams. At the same time, the model (6) is the null hypothesis $\mathcal{H}_0$, assuming that the entries of $\mathbf{z}_i$ do not contribute in the inference of $\mathbf{z}_j$. If $\mathbf{z}_i$ does not cause $\mathbf{z}_j$ the prediction errors of the two models are approximately equal. On the other hand, if $\mathbf{z}_i$ *Granger causes* $\mathbf{z}_j$ the error of model (5) is less than the one of (6). In [17,25] a GC test based on squared sum of residuals is proposed

$$\mathsf{GC}_{i \to j} = \frac{\sum_{t=1}^{T} |\omega_t|^2 - \sum_{t=1}^{T} |\varepsilon_t|^2}{\sum_{t=1}^{T} |\varepsilon_t|^2} \cdot \frac{T - 2K - 1}{K} \overset{\mathcal{H}_1}{\underset{\mathcal{H}_0}{\gtrless}} \theta \tag{7}$$

where $T = K - L$ and K is the length of the time series. Since both the model errors, ε_k and ω_k, follow a normal distribution, the sum of squared residuals is distributed as a central chi-squared. For this reason, the test (7) results in the ratio of chi-squared r.v.'s, giving a $\mathcal{F}$-distribution [25]

$$\mathsf{GC}_{i \to j} \sim \mathcal{F}(L, T - 2L - 1).$$

A fixed false alarm probability, detailed in Sect. 5, determines the threshold θ.

4.2   Transfer Entropy

A criterion to quantify the information flow between two random processes exploiting a variant of conditional mutual information named TE is proposed in [18]. Considering two rows of $\mathbf{Z}$ corresponding to the time series of the transmitted power streams of nodes i and j, respectively, modeled as random processes, the TE from i to j can be expressed as

$$\mathsf{TE}_{i \to j}(R, Q) = \mathcal{I}(z_{j,k}; \mathbf{z}_i^-, \mathbf{z}_j^-) = \mathbb{E}\left\{ \log_2 \frac{p(z_{j,k}|\mathbf{z}_i^-, \mathbf{z}_j^-)}{p(z_{j,k}|\mathbf{z}_j^-)} \right\} \tag{8}$$

where $\mathbf{z}_i^-$ and $\mathbf{z}_j^-$ denote the samples of $\mathbf{z}_i$ and $\mathbf{z}_j$ preceding the time instant k, respectively. Usually, conditional probability densities require the knowledge of infinite past samples of the time series to be evaluated. Anyway, in this formulation, TE is calculated considering only a finite number of past samples for $\mathbf{z}_i$ and $\mathbf{z}_j$ [18]. Thus, the considered time lags for the two time series are given by R and Q, respectively. To obtain the decision threshold θ, the null distribution of the TE has to be approximated using the collected samples [19], choosing a predefined false-alarm probability. Then, the test is

$$\mathsf{TE}_{i \to j} \underset{\mathcal{H}_0}{\overset{\mathcal{H}_1}{\gtrless}} \theta. \tag{9}$$

The response generated by the dataflow from node i to node j might present a delay, i.e., when sending an acknowledgment (ACK). Hence, an interaction delay parameter, n_0, to introduce an offset in the starting point from which the time lag is calculated, is proposed in [29]. The comprehensive definition of TE can be finally written as

$$\mathsf{TE}_{i \to j}(R, Q, n_0) = \mathcal{I}(z_{j,k}; \mathbf{z}_{i,k-n_0-1:k-n_0-R}, \mathbf{z}_{j,k-1:k-Q}). \tag{10}$$

The interaction delay is taken into account only for the time series $\mathbf{z}_i$.

If for a wireless node i, we calculate $\mathsf{TE}_{i \to j}(R, Q, n_0)$, $j = 1, \ldots, N$ with $j \neq i$, and then apply the test described above for each pair $\{i, j\}$, we identify a set of likely neighbours of node i. To lower the number of false neighbors, they can be tested one more time with a slightly different version of TE named conditional transfer entropy (CTE). In this way, the effects of all the detected neighbors on the causal inference are considered. CTE from node i to j is defined as

$$\mathsf{CTE}_{i \to j}(R, Q, n_0, g) = \mathcal{I}(z_{j,k}; \mathbf{z}_{i,k-n_0-1:k-n_0-R}|\mathbf{z}_{j,k-1:k-Q}, \mathbf{z}_{g,k-1:k-Q}) \tag{11}$$

where $g = 1, \ldots, N$ with $g \neq i, j$. For further details about the CTE algorithm please refer to [19].

5   Numerical Results

In this section, the effect of channel impairments on the performance of BSS is evaluated, and the topology sensing methodologies described in Sect. 4 are compared. As a case study, we simulated an IEEE 802.11s ad-hoc network, with channel width of 20 MHz and center frequency $f_0 = 2.412\,\text{GHz}$, on the ns3 network simulator. As mentioned above, the wireless network is located in a square area of side $L_a = 10\,\text{m}$. The transmit power of the nodes of the wireless network is $P_T = 10\,\text{dBm}$, while the thermal noise power at the receivers is $\sigma_N^2 = -93\,\text{dBm}$. The path-loss follows a power-law, with channel gain $h'_{m,n} = h_0(\frac{d_0}{d_{m,n}})^\nu$ where the loss exponent is $\nu = 3$, the reference distance is $d_0 = 1\,\text{m}$, and $h_0 = -60.1\,\text{dB}$. The bandwidth of the RF sensor is $W = 20\,\text{MHz}$, while the integration time is $T_b = 10\,\mu\text{s}$, hence $N_{\text{DOF}} = 2WT_b = 400$. The size of data packets is fixed to 1024 Byte, while the ACKs have a size of 112 Byte. The offered traffic of each node is 1 Mb/s.

The excision filter parameter, λ_n, is set as in (4) with $q = 0.7$. The time lags of the topology inference algorithms are set as $L = 4$, $R = 2$, $Q = 1$, and $n_0 = 3$ according to the Akaike information criterion (AIC) [21]. The false alarm probability for the decision threshold is set to 10^{-2} for both the algorithms. The results presented in this section are obtained from the simulations of $M_{\text{top}} = 100$ different mesh topologies. For each of the simulated topologies, $M_{\text{mc}} = 100$ Monte Carlo trials are performed to randomly deploy the nodes and sensors within the landscape. Usually, the topology of a network is represented by a sparse adjacency matrix $\mathbf{A}$. Thus, the accuracy, a standard non-weighted metric, is not a suitable figure of merit for the evaluation of the topology sensing performance. For this reason, in this work the detection probability (or recall), $p_D^{h,k}$, and the false alarm probability (or false positive rate), $p_{FA}^{h,k}$, of the directed link from node h to node k, defined as

$$p_D^{h,k} = \mathbb{P}\{\hat{a}_{h,k} = 1 | a_{h,k} = 1\}$$
$$p_{FA}^{h,k} = \mathbb{P}\{\hat{a}_{h,k} = 1 | a_{h,k} = 0\},$$

have been adopted. In the following, we refer to p_D and p_{FA} as the detection and false alarm probabilities, respectively, averaged over all the possible network connections.

Table 1. List of the parameters adopted during the simulations.

Parameter Set	A_1	A_2	B_1	B_2	C_1	C_2
ρ_S (nodes/m^2)	0.3	0.3	0.2	0.2	0.1	0.1
σ_S (dB)	3	6	3	6	3	6

The sensed topologies represent the topology of the network in a snapshot taken during the observation time, $T_{ob} = 1\,\text{s}$.

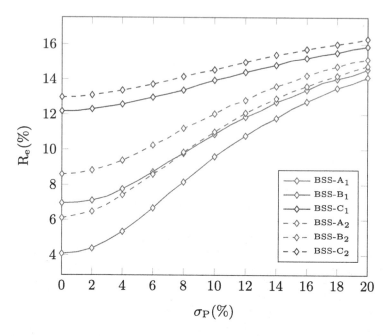

Fig. 2. BSS algorithm performance as a function of the standard deviation σ_P of the uncertainty on the position of the wireless nodes, the shadowing parameter σ_S and the density ρ_S.

5.1   BSS Reconstruction Error

As a figure of merit to quantify the effectiveness of BSS we adopted the reconstruction error, defined as

$$R_e = \frac{\#\,\text{of wrong samples}}{\#\,\text{of total samples}} = \frac{||\mathbf{Z} - \bar{\mathbf{P}}||_1}{N \cdot K}$$

where the matrix $\bar{\mathbf{P}}$ has entries $\bar{p}_{n,k} = 1$ if node n is transmitting in the kth bin, i.e., $p_{n,k} > 0$, and 0 otherwise. Due to the fact that the signals association can be degraded by a non-perfect knowledge of the position of the nodes, we introduce uncertainty by adding a normal distributed r.v. with standard deviation σ_P to the real coordinates. The impact of the number of sensors on the signal separation methods' performance is characterized by the density of sensors ρ_S, defined as the number of sensors per square meter. Table 1 summarizes all the configurations of parameters adopted during the simulations.

The BSS performance tests are shown in Fig. 2. In particular, the proposed method has been tested varying the standard deviation $\sigma_P(\%)$, defined as percentage of the dimension of the scenario, the shadowing parameter σ_S, and the density ρ_S. The figure shows how the reconstruction error rises when increasing σ_P, even at high densities of sensors, i.e., $\rho_S = 0.3$. Moreover, the curves is shifted upward when the shadowing intensity is increased, reaching an error $R_e = 16\%$ with $\sigma_S = 6\,\text{dB}$ and $\sigma_P = 20\%$.

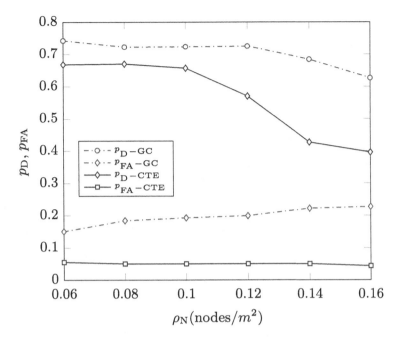

Fig. 3. p_D and p_{FA} of the topology sensing techniques varying the density of nodes ρ_N for $\rho_S = 0.3$ sensors/m^2 and $\sigma_S = 3$ dB.

5.2 Topology Inference and Number of Nodes

To provide a strong comparison between the topology sensing methodologies presented in Sect. 4, the density of wireless nodes per square meter, ρ_N, is varied across the simulations. The BSS that precedes the topology inference was performed with $\rho_S = 0.3$ sensors/m^2, $\sigma_S = 3$ dB and $\sigma_P = 0$. Note that an increase in the number of nodes within the network ·area rises the collision probability, which inevitably translates into a stronger network congestion. The CTE method presents the lower p_{FA}, but the p_D is lower than the GC. This means that the error on the unmixing of the signals impacts more CTE than GC (Fig. 3).

5.3 Impact of Shadowing

In a realistic scenario, the quality of the time series reconstructed after the BSS might be degraded by the presence of shadowing. Such impairment impacts the performance of the algorithms. On this point, it is important to study the accuracy of the algorithms with different propagation characteristics. In Fig. 4, it is shown how an increase in σ_S drops the performance of the algorithms. Here we set the density of sensors $\rho_S = 0.3$ sensors/m^2, the density of nodes $\rho_N = 0.06$ nodes/m^2, and $\sigma_P = 0$. Again, CTE presents a false alarm rate lower than the GC method, but p_D is still the lowest.

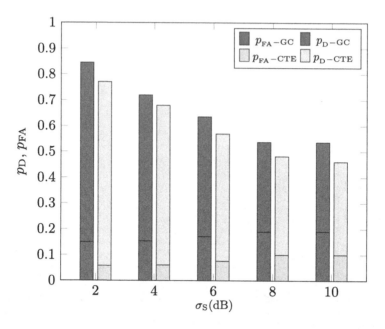

Fig. 4. p_D and p_{FA} of the topology sensing techniques varying the shadowing parameter $\sigma_S(dB)$ for $\rho_S = 0.3$ sensors/m^2.

6   Conclusion

This paper proposed a tool for non-collaborative wireless network topology inference using power profiles captured by radio-frequency (RF) sensors. The framework combines BSS, measurement association, excision filtering, and topology sensing. For the topology sensing, state-of-the-art techniques for causality inference such as Granger causality (GC) and conditional transfer entropy (CTE), that identify causal relationships between the separated traffic profiles, were adopted. The numerical results, accounting for realistic channel impairments (i.e., noise and shadowing) and packet collisions, showed that topology sensing is possible in this framework. As expected, the phenomena mentioned above degrade the performance of the BSS causing a non-perfect reconstruction of the power profile of the transmitted signals. Despite this, the proposed tool senses the topology with promising accuracy when operating in mild shadowing conditions, even with a relatively low number of RF sensors.

References

1. Al-Fuqaha, A., Guizani, M., Mohammadi, M., Aledhari, M., Ayyash, M.: Internet of things: a survey on enabling technologies, protocols, and applications. IEEE Commun. Surv. Tutor. **17**, 2347–2376 (2015)

2. Cichocki, A., et al.: Tensor decompositions for signal processing applications: from two-way to multiway component analysis. IEEE Signal Process. Mag. **32**(2), 145–163 (2015)
3. Cichocki, A., Zdunek, R., Amari, S.: New algorithms for non-negative matrix factorization in applications to blind source separation. In: IEEE International Conference on Acoustics Speech and Signal Processing Proceedings, vol. 5, Toulouse, France, June 2006
4. Favarelli, E., Testi, E., Pucci, L., Giorgetti, A.: Anomaly detection using WiFi signals of opportunity. In: IEEE International Conference on Signal Processing for Communication Systems (ICSPCS), Surfers Paradise, Gold Coast, Australia, December 2019
5. Granger, C.W.J.: Investigating causal relations by econometric models and cross-spectral methods. Econometrica **37**(3), 424–438 (1969)
6. Hyvarinen, A.: Fast and robust fixed-point algorithms for independent component analysis. IEEE Trans. Neural Netw. **10**(3), 626–634 (1999)
7. IEEE 802.22: Standard for Wireless Regional Area Networks-Part 22: Cognitive Wireless RAN Medium Access Control (MAC) and Physical Layer (PHY) specifications: Policies and procedures for operation in the TV Bands, July 2011
8. Ivkovic, G., Spasojevic, P., Seskar, I.: Localization of packet based radio transmitters in space, time, and frequency. In: Asilomar Conference on Signals, Systems and Computers. Pacific Grove, CA, USA, November 2008
9. Joho, M., Mathis, H., Lambert, R.H.: Overdetermined blind source separation: using more sensors than source signals in a noisy mixture. In: International Conference on Independent Component Analysis and Blind Signal Separation (ICA), Helsinki, Finland, pp. 81–86, June 2000
10. Kontos, T., Alyfantis, G.S., Angelopoulos, Y., Hadjiefthymiades, S.: A topology inference algorithm for wireless sensor networks. In: IEEE Symposium on Computers and Communications (ISCC), Cappadocia, Turkey, pp. 479–484, July 2012
11. Laghate, M., Cabric, D.: Learning wireless networks' topologies using asymmetric Granger causality. IEEE J. Sel. Topics Signal Process. **12**(1), 233–247 (2018). https://doi.org/10.1109/JSTSP.2017.2787478
12. Mariani, A., Giorgetti, A., Chiani, M.: Effects of noise power estimation on energy detection for cognitive radio applications. IEEE Trans. Commun. **59**(12), 3410–3420 (2011). https://doi.org/10.1109/TCOMM.2011.102011.100708
13. Mariani, A., Giorgetti, A., Chiani, M.: Model order selection based on information theoretic criteria: design of the penalty. IEEE Trans. Signal Process. **63**(11), 2779–2789 (2015)
14. Nurminen, H., Dashti, M., Piché, R.: A survey on wireless transmitter localization using signal strength measurements. Wirel. Commun. Mobile Comput. **2017**, 1–12 (2017)
15. Pahlavan, K., Li, X., Ylianttila, M., Chana, R., Latva-aho, M.: An overview of wireless indoor geolocation techniques and systems. In: Omidyar, C.G. (ed.) MWCN 2000. LNCS, vol. 1818, pp. 1–13. Springer, Heidelberg (2000). https://doi.org/10.1007/3-540-45494-2_1
16. Pearl, J.: Causality: Models, Reasoning and Inference, 2nd edn. Cambridge University Press, New York (2009)
17. Qin, X., Lee, W.: Statistical causality analysis of INFOSEC alert data. In: Vigna, G., Kruegel, C., Jonsson, E. (eds.) RAID 2003. LNCS, vol. 2820, pp. 73–93. Springer, Heidelberg (2003). https://doi.org/10.1007/978-3-540-45248-5_5
18. Schreiber, T.: Measuring information transfer. Phys. Rev. Lett. **85**, 461–464 (2000). https://doi.org/10.1103/PhysRevLett.85.461

19. Sharma, P., Bucci, D.J., Brahma, S.K., Varshney, P.K.: Communication network topology inference via transfer entropy. IEEE Trans. Netw. Sci. Eng. **7**, 1–7 (2019). https://doi.org/10.1109/TNSE.2018.2889454

20. Sithamparanathan, K., Giorgetti, A.: Cognitive Radio Techniques: Spectrum Sensing, Interference Mitigation and Localization. Artech House Publishers, Boston (2012)

21. Stoica, P., Selen, Y.: Model-order selection: a review of information criterion rules. IEEE Signal Proc. Mag. **21**(4), 36–47 (2004)

22. Testi, E., Favarelli, E., Giorgetti, A.: Machine learning for user traffic classification in wireless systems. In: European Signal Processing Conference (EUSIPCO), Rome, Italy, pp. 2040–2044, September 2018. https://doi.org/10.23919/EUSIPCO.2018.8553196

23. Testi, E., Favarelli, E., Pucci, L., Giorgetti, A.: Machine learning for wireless network topology inference. In: International Conference on Signal Processing and Communication Systems (ICSPCS), Surfers Paradise, Gold Coast, Australia, December 2019

24. Testi, E., Giorgetti, A.: Blind wireless network topology inference. IEEE Trans. Commun. **69**, 1109–1120 (2020)

25. Tilghman, P., Rosenbluth, D.: Inferring wireless communications links and network topology from externals using Granger causality. In: IEEE MILCOM, Military Communications Conference (MILCOM), San Diego, CA, USA, pp. 1284–1289, November 2013. https://doi.org/10.1109/MILCOM.2013.219

26. Urkowitz, H.: Energy detection of unknown deterministic signals. Proc. IEEE **55**(4), 523–531 (1967)

27. Wax, M., Kailath, T.: Detection of signals by information theoretic criteria. IEEE Trans. Acoust. Speech Signal Process **33**(2), 387–392 (1985)

28. Wenli, J., Teng, G., Meiyin, J.: Researching topology inference based on end-to-end date in wireless sensor networks. In: International Conference on Intelligent Computation Technology and Automation (ICICTA), Shenzhen, China, vol. 2, pp. 683–686, April 2011

29. Wibral, M., et al.: Measuring information-transfer delays. Plos One **8**, 1–19 (2013)

30. Xu, H., Farajtabar, M., Zha, H.: Learning Granger causality for Hawkes processes. In: International Conference on Machine Learning ICML, New York, NY, USA, vol. 48, February 2016

31. Yu, C., Chen, K., Cheng, S.: Cognitive radio network tomography. IEEE Trans. on Veh. Technol. **59**(4), 1980–1997 (2010)

32. Yu, X., Hu, D., Xu, J.: Blind Source Separation: Theory and Applications, 1st edn. Wiley, New York (2014)

33. Vardi, Y.: Network tomography: estimating source-destination traffic intensities from link data. J. Am. Stat. Assoc. **91**(433), 365–377 (1996)

Resource Management
and Optimization

Efficient Clustering Schemes Towards Information Collection

Krzysztof Cichoń$^{(\boxtimes)}$ and Adrian Kliks

Institute of Radiocommunications, Poznan University of Technology, Poznan, Poland
{krzysztof.cichon,adrian.kliks}@put.poznan.pl

Abstract. One of the challenges in cooperative spectrum sensing is to optimize the energy consumption of the network. Delivery of all measurements from all the nodes to the fusion centre is not the best solution from the perspective of energy-efficiency. Clustering of nodes with similar channel conditions may reduce the amount of transmitted data, and in consequence reduce the amount of consumed energy. In this paper we investigate the performance of selected algorithms known in the domain of artificial intelligence, applied to perform reliable yet energy-aware spectrum sensing.

Keywords: Signal detection · Clustering · Classification

1 Introduction

One of the key assumptions in the cognitive radio networks (CRN) is to sufficiently protect the primary user (PU) transmissions while utilizing the spectrum resources in the flexible way. Operational wireless systems deployed over certain area cannot be disrupted by the secondary user (SU), at most some small and acceptable amount of interference power can be induced [1]. In consequence, the SU has to be aware about the presence of the existing transmissions, and one way to achieve it is to observe - *sense* - the spectrum in order to detect any human-made signals.

Since the origin of the cognitive radio technology (over two decades ago [3]), numerous papers have been published, where researchers investigated various aspects of the spectrum sensing procedure [2]. Interesting survey on the spectrum sensing techniques may be found in [4,5], where the authors reviewed the performance of numerous algorithms designed to spectrum occupancy detection. However, it has been stated that the single-node spectrum sensing (i.e., the procedure where the SU observes spectrum solely without any help or information exchange with other terminals or sensors) cannot guarantee high reliability and accuracy regardless of observation time, as such an approach is vulnerable to e.g. hidden node problem [6]. In this scenario, given SU terminal will be not able to detect the presence of nearby PU due to high shadowing of the signal due to the presence of obstacle (e.g. building). In consequence, single-node spectrum sensing has to be either complemented with the information delivered from the

G. Caso et al. (Eds.): CrownCom 2020, LNICST 374, pp. 45–58, 2021.
https://doi.org/10.1007/978-3-030-73423-7_4

dedicated geolocation databases or collaborative (cooperative) solutions have to be considered. In the latter case even great number of nodes may exchange information to better detect the presence of PU [7]. However, the presence and cooperation between numerous sensors in the system entails the increased energy consumption by the whole system, as well as the increase in latency in decision making. It is due to the need for often periodic delivery of measurement updates from all or most of the cooperating sensors. There is then a trade-off between the number of nodes used to increase the awareness about the surrounding radio environment, and the consumed energy and introduced delay. As sensing is an inherent feature of SU node, the overall energy optimization may be achieved by proper grouping of sensing nodes. Within such a group, the information is processed locally (e.g. the local decision on the spectrum occupancy is made) and later delivered by the group leader to the dedicated sink-node in the network, called often a fusion centre (FC). The FC node collects the messages delivered from all associated clusters, analyse them, perform decisions and distribute them among these clusters.

However, in practice, the locations of the sensing nodes (including wireless user terminals) is not known a-priori, and has to be learnt. It means that the clusters of SU terminals have to be created dynamically, in autonomous and distributed way, as well as the cluster-heads have to be selected in similar way. Clustering of nodes is widely applied and well investigated research problem present in various domains of wireless communications, e.g. [8,9]. In advanced scenarios, where e.g. the reliability of sensed data is not stable, the algorithms known from the domain of artificial intelligence (AI) can be applied. In particular, the clustering algorithms may be considered as efficient tools for node grouping in CRN. Being an extension of the previous paper [10], in this work we compare the proposed soft-K-means algorithms with other AI solutions, properly adjusted to the investigated scenario and applied system model. The rest of the paper is organized as follows. In Sect. 2, the system model and the research problem is presented, and the basics of cooperative spectrum sensing is overviewed. In Sect. 3, the details of considered clustering algorithms are described. Section 4 deals with results analysis, and the whole work is concluded in Sect. 5.

2   Energy Efficiency in Cooperative Spectrum Sensing

In our scheme we consider the set of N wireless nodes capable in performing spectrum sensing in a cooperative way. In order to optimize the energy consumption in the whole network, dynamic node clustering is considered, where cluster representatives (cluster heads, CH) are selected in a dynamic way. Once all the CH are identified, they communicate with the dedicated central node, FC, in order to deliver to it all local information and allow it to make reliable global decision about the spectrum occupancy. As discussed later, the CHs may either send the unprocessed (raw) data gathered from all nodes belonging to the cluster, slightly processed data or even the local decisions. Without loss of generality, we assume that the messages between the nodes are transmitted applying simple binary modulation scheme, i.e. binary phase shift keying (BPSK). The considered scenario is illustrated in Fig. 1.

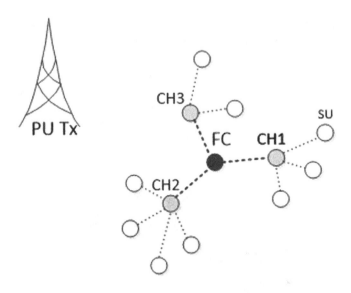

Fig. 1. System model where SUs are sensing PU's signal and are clustered into three clusters (PU - Primary User, SU - Secondary User, CH - Cluster Head, FC - Fusion Centre)

The ultimate goal of any spectrum sensing algorithm is to correctly detect the presence or absence of the signal in the observed band at the specific location. Numerous single-node spectrum sensing solutions have been proposed in the rich literature, starting from the simple yet highly inaccurate in poor channel conditions algorithm known as energy detection, through more advanced eigenvalue detectors or covariance based algorithms, finishing at highly complicated matched-filter schemes [4,5]. However, it has been proved that none of these solutions can guarantee high performance in specific yet common situation, e.g., in case of hidden node problem. In consequence, single node spectrum sensing is said to be not reliable enough to allow for secondary transmission is real-world applications. Contrarily, when the neighbouring nodes could communicate and exchange information, the level of the network awareness on the surrounding radio environment increases.

There are four cases that may happen when the node performs sensing - and all of them constitutes the so-called confusion matrix depicted in Fig. 2. When the node detects the signal if it is indeed presence, one may speak of probability of detection or true-positive scheme. Next, when the channel vacancy is decided when there is no real signal, one may thing on negative-positive scheme. Finally, there are two kinds of wrong decisions - classified as miss-detection (when the real signal is present but is not detected by the SU, false negative case) and classified as probability of false-alarm - the detector decides on signal presence where in reality there is none (false positive case). This decision making process is often referred to hypothesis testing, where absence of PU signal is treated as the null-hypothesis [11,12].

In cooperative case, the nodes transmit their data to the selected cluster head, which is responsible for their processing and final decision making. The nodes may transmit raw data or some initial decisions to the cluster head, which in turns transmit a joint message to the global fusion centre (one dedicated node) to make global decision. In other approach, once the cluster representative node is selected for a certain group of nodes, only this node is responsible for delivering sensing information to the FC. When the fusion centre receives all messages from associated sensing nodes, it may perform final decision by merging delivered information, applying one of the possible rules (OR, AND or K-out-of-N [7]).

		Actual state	
		PU present	PU absent
Detector decision	PU present	Correct positive detection	False alarm
	PU absent	Miss-detection	Correct negative decision

Fig. 2. Confusion matrix of the predicted and actual state in spectrum sensing

Thus, also from this perspective it is reasonable to consider node grouping. Various criteria may be applied in this respect, for example, one may propose to group together the nodes which are geographically close to each other and create clusters based on density of nodes in various areas. Another approach is to group all nodes with similar channel conditions. These issue will be discussed in detail in the following section. Moreover, from the perspective of energy consumption in the CRN, when the total number of cooperating nodes increases, the whole consumed energy increases as well. One may observe that in general, the longer the distance between the nodes, the greater the transmit power necessary to deliver the message to the destination node, and in consequence greater the total energy consumption. In order to evaluate the energy-efficient solutions in CRN there is a need for proper power consumption models, as presented in [10]. The total power consumed by N network members is given by the following formula:

$$P_{\text{total}} = \sum_{i=1}^{N} \left(P_{\text{sens}}^{(i)} + P_{\text{rep}}^{(i)} \right) + P_{\text{shar}}, \tag{1}$$

where $P_{\text{sens}}^{(i)}$ is the power related to spectrum sensing for SU i, which is a constant value for the selected sensing technique and depends on the complexity of the sensing technique. P_{shar} is the power devoted to information sharing within the network, $P_{\text{rep}}^{(i)}$ is the power related to reporting of SU's observation to the cluster head by node i or the power of reporting by the cluster head to Fusion Centre. Reporting power $P_{\text{rep}}^{(i)}$ depends on the multipath loss $\eta^{(i)}$, as well as on the distance between the i-th node and cluster head $d_{\text{CH}}^{(i)}$:

$$P_{\text{rep}}^{(i)} = \frac{\gamma_{target}^{\text{BPSK}} \left(d_{\text{CH}}^{(i)}\right)^{\xi} \sigma_n^2}{\eta^{(i)} G_{\text{t}} G_{\text{r}} \left(\frac{\lambda}{4\pi}\right)^2}, \tag{2}$$

where $\gamma_{target}^{\text{BPSK}}$ is the target signal-to-noise ratio (SNR) for a BPSK-modulated message, ξ is an exponent of the received power decrease dependent on the type of wireless environment, σ_n^2 is the power of noise, G_{t} and G_{r} are the gains of the transmitting and receiving antennas, respectively, and $\frac{\lambda}{4\pi}$ is the wave number [13].

Knowing the energy consumption model, one may formulate the investigation problem, i.e., how to properly cluster the nodes based on the similarity of the observed channel conditions in order to guarantee high energy-efficiency of the entire network. Various algorithms may be applied here, as discussed in the following section.

3   Clustering Algorithms

As specified in the previous sections, we consider the scheme where every SU collects information from the surrounding environment, and delivers it to the cluster head. However, from the network perspective the number of possible clusters as well as the selection of the cluster leader have great impact on the overall energy efficiency. Various clustering criteria may be identified, such as grouping nodes which are close to each other in Euclidean sense. In our scheme, we extend this approach by clustering the nodes which observe similar channel conditions. In other words, when there is a set of nodes even closely deployed, but portion of them is subject to some strong shadowing (with respect to the fusion centre), at least two clusters will be created - one for shadowed and one for non-shadowed group of SUs. In our study we compare some AI based solutions for node clustering (see e.g. [14]). The exemplary clustering results are presented for all algorithms in Fig. 3.

Standard K-means. K-means is the well-known clustering scheme in which the target number of clusters K has to be known before the start of the algorithm [15]. In short, the algorithm operates in iterative mode - in the first *assignment* step, each SU is assigned to the closest centroid (cluster centre). In second step the positions of centroids are updated on the base of assigned cluster members. Thus, the second phase is called an *update step*.

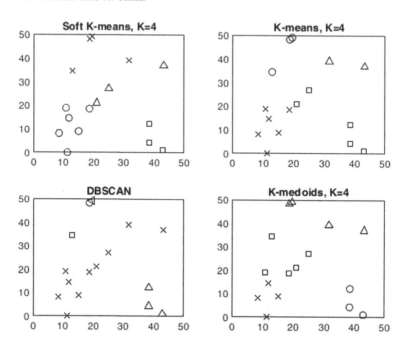

Fig. 3. Exemplary clustering results for $N = 16$ SUs for clustering methods considered in the article. The same node distribution is applied to find the members of clusters (distinguished with various shapes: squares, triangles, circles, etc.)

Soft K-means. In the soft K-means the way how the nodes are assigned to clusters is *soft* what means that the node can be *partially* member of more than one cluster. The *degree of assignment* is then given by the formula:

$$r_k^{(n)} = \frac{\exp\left(-\theta d_{m^{(k)}}^{(i)}\right)}{\sum_{k'} \exp\left(-\theta d_{m^{(k')}}^{(i)}\right)}, \tag{3}$$

where $d_{m^{(k)}}^{(i)}$ is the distance between SU i and the centroid $m^{(k)}$ of cluster k. The sum of all assignments for the nth SU to all centroids considered in the network is one. θ is the stiffness which is the key parameter for the algorithm and is an inverse-length-squared.

K-medoids. In K-medoids the clustering is considered to be similar to K-means article. In the latter, the new centroid is found as the mean of locations of cluster members. The algorithm may be vulnerable to outlier nodes which are distant from main cluster members and may disturb the correct location of cluster head. However, different than in K-means, the location of the medoid is always the position of the real node, since each SU is tested as the potential medoid. Moreover, in K-medoids the central point of cluster is selected on the base of the closest distance to remaining potential cluster members. The procedure is iteratively conducted for any network member.

Density-Based Approach. Density-based spatial clustering, known as DBS-CAN, is the algorithm where the number of clusters is not needed to be specified at the beginning of the algorithm. Instead, two other parameters are required: ϵ, being the radius of the neighbourhood under consideration, and minimum cluster size. First, the neighbours are found within the ϵ neighbourhood. If the number of neighbours is lower than minimum cluster size, then these nodes become so-called *noise nodes*, otherwise they form the cluster and become *clustered nodes*. The cluster could be extended after the time the procedure is repeated for border members of newly formed cluster. If no new cluster members could be added, the procedure is repeated for other unlabelled nodes with the possible formation of different cluster. The key parameter in the algorithm is the selection of the ϵ value which is the radius around the point which is used for adding new cluster members. Moreover, the minimum cluster size is the second important parameter which should be no lower than 3.

4   Simulation Results

Extensive simulations have been carried out toward reliable performance comparison of the selected well-known clustering algorithms, i.e., K-means, K-medoids and DBSCAN with the proposed soft-K-means solution [10]. In our simulation we consider the set of N from 8 to 20 nodes deployed uniformly over the considered geographical area of size 50 m. The nodes can communicate wirelessly by sending BPSK modulated signals. The required SNR that has to be guaranteed at the receiving nodes was set to $\gamma_{target}^{BPSK} = 11.3$ which is found for target BER equal to 10^{-6}. The value of the channel decaying factor ξ was equal to 3.5. The transmit and receive antenna gains have been setup to $G_t = 1$ and $G_r = 1$, respectively, and the considered wavelength was found for carrier frequency of 2.4 GHz $\lambda = \frac{c}{f_c} = 0.125$ m.

In the presented results two reference cases without clustering are considered. In the first one, called as 'SNR-based selection', a majority of nodes report their observation to FC, the nodes with the highest SNR are selected. In 'energy-based selection' the same number of nodes is selected, based on the lowest energy usage in reporting link. In the following section the 'reporting node' is the node which reports sensing observations directly to FC, regardless of the clustering procedure is applied (DBSCAN, K-means, K-medoids) or not (SNR- and energy based selections).

First, DBSCAN algorithm is investigated for finding such set of parameters which guarantee efficient operation. In Fig. 4, the consumed energy of the network is presented as the function of the number of nodes. One can observe that the minimum energy (given in Joules) is observed for various DBSCAN algorithms. First, the consumed energy increases with the number of nodes for any variant of DBSCAN algorithm. And for 8–10 nodes the lowest energy usage is for $\epsilon = 16$, but this case for the network of 20 nodes becomes inefficient. Then the trend is the following: the greater the number of nodes in the network, the lower the optimum value of ϵ: for example for 14–18 nodes the lowest energy usage is observed for $\epsilon = 12$ and for 20 nodes for $\epsilon = 10$. The reference cases, i.e. energy selection and SNR-based selection (for significant number of nodes), offer much greater energy consumption than DBSCAN grouping.

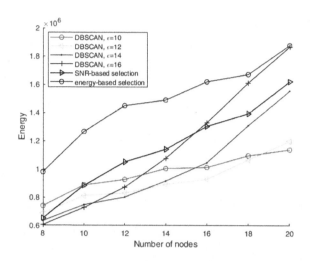

Fig. 4. Energy usage by DBSCAN algorithm with various values ϵ

This is mainly due to the greater energy consumption for reporting links from SUs to Fusion Centre than to cluster heads. In Fig. 5 one can see the number of reporting nodes which for reference cases is significantly greater than for DBSCAN algorithms. This is because in reference cases clusters are not formed, the SUs are independent and report observation directly to Fusion Centre. However, in DBSCAN parameter ϵ discriminates the number of reporting nodes (which is the number of clusters and *noise* nodes. With the increasing number of nodes the number of reporting links is slightly going down. The reason for this is that the greater number of nodes, the lower number of possible *noise* nodes and it is more probable to form the cluster with minimum number of nodes required in the algorithm.

In Fig. 6 the energy usage for K-medoids is shown for various number of clusters K. The lowest energy consumption is observed for two cases: for 4 and 5 clusters. The energy usage raises with number of nodes but is still significantly lower than for reference cases (with the exception of SNR-based scheme for low number of nodes).

Then, in Fig. 7 one can see the energy usage for standard K-means scheme. The energy usage is similar within a couple of cases where the number of clusters is from 3 to 5. The case with $K = 2$ clusters offer higher energy usage what makes it unuseful.

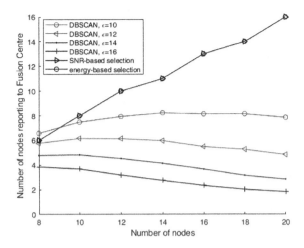

Fig. 5. Number of reporting nodes for DBSCAN algorithm with various values ϵ

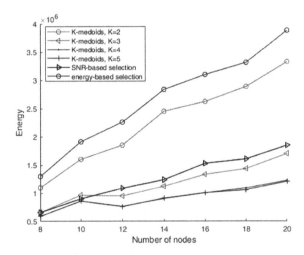

Fig. 6. Energy consumption for the whole network within K-medoids algorithm

The comparison of energy consumption for soft K-means is highlighted in Fig. 7. The crucial parameter of *stiffness* is set for all cases as 0.15 according to the highlights in [10]. However, energy usage for three cases is similar: $K = \{3, 4, 5\}$. Slightly lower energy usage is observed for $K = 4$ clusters. What is important in soft K-means the energy spent in the sensing process is lower than in standard K-means (Fig. 8).

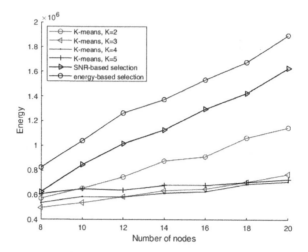

Fig. 7. Energy consumption for the whole network within standard K-means

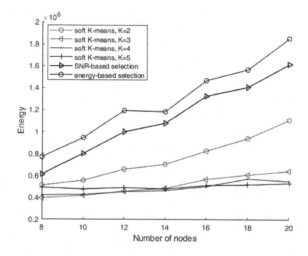

Fig. 8. Energy consumption for the whole network within soft K-means

Finally, the results for all considered algorithms were compared in Fig. 9. The input parameters were selected in the way that 4 clusters is found. In DBSCAN, where the number of clusters cannot be defined a priori since it is the result of ϵ and minimum cluster size parameters, it is around 4 for $\epsilon = 14$. Within such set of algorithms the lowest energy usage is observed for soft K-means. The energy usage is of 0.5 J and increases slightly when more nodes are considered. The second-best is standard K-means offering also good stability of energy consumption versus number of nodes and offering about 35% higher energy usage than soft K-means. K-medoids and DBSCAN offer similar performance to standard K-means when number of nodes is of 8 to 12 nodes. However, for greater network size the consumed energy is far bigger than for K-means schemes. Moreover, the reference cases, based on simple centralised topologies involve significantly higher energy usage than clustered schemes.

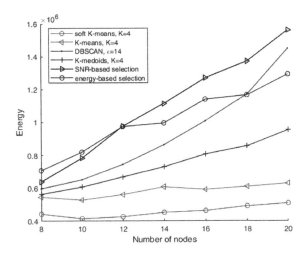

Fig. 9. Energy consumption comparison for all considered grouping algorithms

The reason for this is much greater number of reporting nodes when compared to other grouping schemes, what is included in Fig. 10. Indeed, K-medoids and both versions of K-means have constant number of 4 reporting nodes, in DBSCAN it decreases from 5 to 3, however, in SNR and energy selection number of reporting nodes is the total number of network members and increases from 6 to 16 (around 75% of network members). Therefore, the soft K-means is the most promising solution since due to its soft *stiffness* metric the required number of K clusters may be lowered by merging some clusters according to the procedure. This can be observed in Fig. 10 where the number of reporting nodes for soft K-means is lower than 4. Moreover, it offers the same detection quality (Fig. 11) with lower energy consumption when compared to other clustering solutions. The global detection quality for all clustering schemes remains the same for a given number of nodes and grows when network size increases.

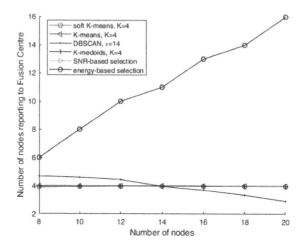

Fig. 10. Number of clusters formed for various clustering schemes

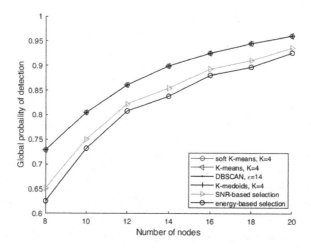

Fig. 11. Global probability of detection for various clustering schemes

5   Conclusions

In this work the analysis of energy efficiency for selected clustering procedures is provided. First, density-based approach with noise nodes is analysed and optimum values for input parameters are found. Then, the energy usage for K-medoids, K-means and proposed soft K-means is validated. The conducted simulations have shown the lowest energy usage is observed for soft K-means scheme. What is more, the detection quality is unchanged. Then, for K-means, the energy usage is higher, but low complexity is a great advantage. Moreover soft and standard K-means both offer the energy usage which is not growing rapidly with number of nodes in the network. In future works it is planned to compare also distributed clustering schemes with the presented hierarchical ones. Moreover, mobility of SUs and network is the important factor for detection quality and energy usage and is planned to be included in the model.

Acknowledgement. The presented work has been funded by the National Science Centre in Poland within the CERTAIN project (no. 2017/27/L/ST7/03166) of the DAINA programme.

References

1. Haykin, S.: Cognitive radio: brain-empowered wireless communications. IEEE J. Sel. Areas Commun. **23**(2), 201–220 (2005). https://doi.org/10.1109/JSAC.2004. 839380
2. Haykin, S., Thomson, D.J., Reed, J.H.: Spectrum sensing for cognitive radio. Proc. IEEE **97**(5), 849–877 (2009). https://doi.org/10.1109/JPROC.2009.2015711
3. Mitola, J., Maguire, G.Q.: Cognitive radio: making software radios more personal. IEEE Pers. Commun. **6**(4), 13–18 (1999). https://doi.org/10.1109/98.788210
4. Yucek, T., Arslan, H.: A survey of spectrum sensing algorithms for cognitive radio applications. IEEE Commun. Surv. Tutor. **11**(1), 116–130 (2009). https://doi.org/ 10.1109/SURV.2009.090109
5. Ali, A., Hamouda, W.: Advances on spectrum sensing for cognitive radio networks: theory and applications. IEEE Commun. Surv. Tutor. **19**(2), 1277–1304 (2017). https://doi.org/10.1109/COMST.2016.2631080
6. Nekovee, M.: A survey of cognitive radio access to TV white spaces. In: 2009 International Conference on Ultra Modern Telecommunications and Workshops, St. Petersburg, pp. 1–8 (2009). https://doi.org/10.1109/ICUMT.2009.5345318
7. Cichoń, K., Kliks, A., Bogucka, H.: Energy-efficient cooperative spectrum sensing: a survey. IEEE Commun. Surv. Tutor. **18**(3), 1861–1886 (2016). https://doi.org/ 10.1109/COMST.2016.2553178
8. Din, S., Ahmad, A., Paul, A., Ullah Rathore, M.M., Jeon, G.: A cluster-based data fusion technique to analyze big data in wireless multi-sensor system. IEEE Access **5**, 5069–5083 (2017). https://doi.org/10.1109/ACCESS.2017.2679207
9. Zhang, D., Ge, H., Zhang, T., Cui, Y., Liu, X., Mao, G.: New multi-hop clustering algorithm for vehicular ad hoc networks. IEEE Trans. Intell. Transp. Syst. **20**(4), 1517–1530 (2019). https://doi.org/10.1109/TITS.2018.2853165
10. Cichoń, K.: Soft network organisation towards future distributed ML-Based sensing systems. In: 28th International Conference on Software, Telecommunications and Computer Networks, Hvar (2020). https://doi.org/10.23919/SoftCOM50211.2020. 9238189
11. Jayaweera, S.K.: Introduction to detection theory. In: Signal Processing for Cognitive Radios, pp. 65–131. Wiley (2015). https://doi.org/10.1002/9781118824818. ch4
12. Hippenstiel, R.D.: Detection Theory: Applications and Digital Signal Processing, 1st edn. CRC Press, Boca Raton (2001). SBN-10 0849304342. ISBN-13 978-0849304347

13. Cichoń, K., Bogucka, H.: An energy-efficient cooperative spectrum sensing. In: IEEE International Black Sea Conference on Communications and Networking (BlackSeaCom), pp. 24–28 (2015). https://doi.org/10.1109/BlackSeaCom.2015.7185079

14. Xu, R., Wunsch, D.: Clustering. Wiley-IEEE Press, Hoboken (2008). ISBN-10 0470276800. ISBN-13 978-0470276808

15. MacKay, D.J.C.: Information Theory, Inference, and Learning Algorithms. Cambridge University Press, Cambridge (2002)

A Non-zero Sum Power Control Game with Uncertainty

Andrey Garnaev[✉]

WINLAB, Rutgers University, North Brunswick, USA

Abstract. We consider the communication between a transmitter (user) and a receiver in the presence of a jammer where the jammer, in contrast to the user, has access to local information about jamming fading gain (reflecting distance of the jammer to the receiver) and jamming cost (reflecting its technical characteristics). The problem is modeled as a Bayesian game. Signal-to-interference-plus-noise ratio (SINR) is considered as user's communication utility. Nash equilibrium as well as Stackelberg equilibrium are derived in closed form and compared.

Keywords: Jamming · Bayesian equilibrium · Incomplete information

1 Introduction

Due to the shared and open-access nature of the wireless medium, wireless networks are vulnerable to jamming attacks. Non-cooperative game theory is natural tools to study such jamming problems [11] due to such problems involve multiple agents (say, a user and a jammer) and each of them has own objective. A key characteristic of wireless access networks is that the agents might not have complete information regarding the other agents' identity, channel characteristics, or location [15]. Typically, in literature, uncertainty on one parameter is considered. For example, in [3,5,6,9], uncertainty about rival's type was investigated. In [12], uncertainty on fading channel gains was considered in the context of single carrier communications with non-hostile interference caused by selfish users, while, in [2,10], with hostile interference caused by an adversary. In [16], uncertainty about the jammer's location was examined within a frequency-division multiple access (FDMA) communication scenario.

In [13], it was suggested to consider combined uncertainty about several parameters, namely, fading gains and transmission cost. Due to that, in [13], the agents do not have access to local information it did not lead to an increase in the number of strategies compared with the complete information scenarios.

In this paper, we fill the gap in the literature on the case when one of the agent, namely, the user, does not have access to combined local information about the gain of the jammer's channel (reflecting the distance from the jammer to the receiver) and the jamming cost (reflecting technical characteristics of the jammer). While the jammer has access to such local information. This corresponds the most dangerous scenarios of the user's communication with the

© ICST Institute for Computer Sciences, Social Informatics and Telecommunications Engineering 2021
Published by Springer Nature Switzerland AG 2021. All Rights Reserved
G. Caso et al. (Eds.): CrownCom 2020, LNICST 374, pp. 59–68, 2021.
https://doi.org/10.1007/978-3-030-73423-7_5

receiver. We consider SINR as user's communication utility. This work is a complementary work to [10], where throughput was considered as user's communication utility and the user has incomplete information only about one network parameter, namely, the jammer's channel gain.

2   Communication Model

In this paper we consider a single carrier communication between a transmitter (user) and a receiver in the presence of a jammer, who intends to degrade the user's communication by generating interference. The channel is assumed to be flat fading. As examples of studying single carrier communication, see, for example, [1,4,7,8,10,13,14,17–20]. We assume that the user does not have access to local information about the jammer's channel gain (reflecting its distance to the receiver) and jamming power cost (reflecting technical characteristics of the jammer). While the jammer has access to such local information. Namely, the user knows that the jammer's channel gain can be g_i with a priori probability α_i where $i \in \mathcal{N} \triangleq \{1,\ldots,n\}$ and $\sum_{i=1}^{n} \alpha_i = 1$, and the (normalized) jamming power cost can be $C_{J,j}$ with a priori probability β_j, for $j \in \mathcal{M} \triangleq \{1,\ldots,m\}$, where $\sum_{j=1}^{m} \beta_j = 1$. The strategy for the user is its transmission power P, with $P \geq 0$. We say that the jammer is of type-(i,j) if its channel gain and the jamming cost are g_i and $C_{J,j}$, respectively. The strategy for the (i,j)-type jammer is its jamming power $J_{i,j}$, with $J_{i,j} \geq 0$. Let[1] $\boldsymbol{J} = (J_{1,1},\ldots,J_{1,m},\ldots,J_{n,1},\ldots,J_{n,m})$. Each type jammer as well as the user have complete knowledge about all possible network parameters. But, in contrast to the jammer, the user does not know what type of the jammer occurs. We consider SINR as the user's communication utility. Let the payoff to the user is the difference between the expected SINR and transmission power cost, while the payoff to the (i,j)-type jammer is the negative of sum of the SINR and jamming power cost:

$$v_U(P, \boldsymbol{J}) = \sum_{(i,j)\in\mathcal{N}\times\mathcal{M}} \alpha_i \beta_j hP/(\sigma^2 + g_i J_{i,j}) - C_P P, \qquad (1)$$

$$v_{J,i,j}(P, J_{i,j}) = -hP/(\sigma^2 + g_i J_{i,j}) - C_{J,j} J_{i,j}, \ (i,j) \in \mathcal{N} \times \mathcal{M}, \qquad (2)$$

where C_P is (normalized) transmission power cost and σ^2 is the background noise.

The user and each type jammer want to maximize their own payoffs. Thus, we look for Nash equilibrium (NE) [11]. Recall that $(P, \boldsymbol{J})$ is an NE if and only if

$$v_U(\tilde{P}, \boldsymbol{J}) \leq v_U(P, \boldsymbol{J}), \forall \tilde{P} \geq 0, \qquad (3)$$

$$v_{J,i,j}(P, \tilde{J}_{i,j}) \leq v_{J,i,j}(P, J_{i,j}), \forall \tilde{J}_{i,j} \geq 0 \text{ and } (i,j) \in \mathcal{N} \times \mathcal{M}. \qquad (4)$$

Denote this game by Γ_N. Note that, in game Γ_N there is at least one NE since $v_U(P, \boldsymbol{J})$ is linear in P and $v_{J,i,j}(P, J_{i,j})$ is concave in $J_{i,j}$ [12].

---

[1] We use bold face font to denote vectors.

In the following sections we derive the NE in closed form via solving the best response equations. By (3) and (4), $(P, \boldsymbol{J})$ is an NE if and only if P is the best response to $\boldsymbol{J}$, while $J_{i,j}$ is the best response to P for each (i, j), i.e., they are solution of the following best response equations:

$$P = \mathrm{BR}_U(\boldsymbol{J}) \triangleq \underset{P \geq 0}{\operatorname{argmax}}\, v_U(P, \boldsymbol{J}), \tag{5}$$

$$J_{i,j} = \mathrm{BR}_{J,i,j}(P) \triangleq \underset{J_{i,j} \geq 0}{\operatorname{argmax}}\, v_{J,i,j}(P, J_{i,j}), \ (i, j) \in \mathcal{N} \times \mathcal{M}. \tag{6}$$

2.1 Auxiliary Notations and Results

Let us introduce auxiliary notations and results employed in the next section to derive in closed form. First let us split the set $\mathcal{N} \times \mathcal{M}$ into subsets $\mathcal{S}_r$, $r = 1, ..., R$ with equal $C_{J,j}/g_i$. Specifically, let

$$\mathcal{N} \times \mathcal{M} = \cup_{r=1}^{R} \mathcal{S}_r \text{ with } C_{J,j}/g_i = C_{J,\tilde{j}}/g_{\tilde{i}}, \forall (i, j) \in \mathcal{S}_r \text{ and } (\tilde{i}, \tilde{j}) \in \mathcal{S}_r. \tag{7}$$

Thus, $C_{J,j}/g_i \neq C_{J,\tilde{j}}/g_{\tilde{i}}$ for any $(i, j) \in \mathcal{S}_r$ and $(\tilde{i}, \tilde{j}) \in \mathcal{S}_{\tilde{r}}$ with $r \neq \tilde{r}$. Let

$$\gamma_r \triangleq \sum_{(i,j) \in \mathcal{S}_r} \alpha_i \beta_j. \tag{8}$$

Motivated by (7), we can introduce the following notation:

$$\overline{C}_r \triangleq C_{J,j}/g_i \text{ for any } (i, j) \in \mathcal{S}_r. \tag{9}$$

Without loss of generality we can assume that the sets $\mathcal{S}_r$ are arranged in increasing order by $\overline{C}_r$, i.e.,

$$\overline{C}_1 < \overline{C}_2 < ... < \overline{C}_R, \tag{10}$$

and let $\overline{C}_0 \triangleq 0$ and $\overline{C}_{R+1} \triangleq \infty$.

Let

$$P_r \triangleq \sigma^4 \overline{C}_r/h \text{ for } r = 0, ..., R+1. \tag{11}$$

Thus, by (10), we have that

$$0 = P_0 < P_1 < P_2 < ... < P_R < P_{R+1} = \infty. \tag{12}$$

By (12), for each $P > 0$ there is an unique integer $r(P) \in \{0, ..., R\}$ such that

$$P_{r(P)} \leq P < P_{r(P)+1}. \tag{13}$$

Thus, we can introduce the following notation:

$$F(P) \triangleq F_{r(P)}(P) \text{ with } P_{r(P)} \leq P < P_{r(P)+1}, \tag{14}$$

where

$$F_r(P) \triangleq \sum_{i=1}^{r} \gamma_i \sqrt{h\overline{C}_i/P} + \frac{h}{\sigma^2} \sum_{i=r+1}^{R} \gamma_i. \tag{15}$$

In the following proposition auxiliary properties of $F(P)$ are given.

Proposition 1. *(a) $F(P)$ is continuous and decreasing from $F(0) = h/\sigma^2$ to zero as $P \uparrow \infty$.*
(b) Let $h/\sigma^2 > C_P$. Then there is an unique r_* such that

$$\chi_{r_*^N+1} < C_P \le \chi_{r_*^N}, \tag{16}$$

where

$$\chi_r \triangleq \frac{h}{\sigma^2} \left(\sum_{i=1}^{r} \gamma_i \sqrt{\overline{C}_i/\overline{C}_r} + \sum_{i=r+1}^{R} \gamma_i \right), \quad r = 1, \ldots, R \tag{17}$$

and $\chi_{R+1} \triangleq 0$.
(c) The following equation has the unique positive root

$$F(P) = C_P. \tag{18}$$

Moreover, this root P is equal to P_N, where

$$P_N \triangleq h \left(\frac{\displaystyle\sum_{i=1}^{r_*^N} \gamma_i \sqrt{\overline{C}_i}}{C_P - (h/\sigma^2) \displaystyle\sum_{i=r_*^N+1}^{R} \gamma_i} \right)^2. \tag{19}$$

Proof: (a) Note that, by (11), $h/\sigma^2 = \sqrt{h\overline{C}_r/P_r}$ for any r. This and (14) imply that $F(P)$ is continuous. Also, $F(P)$ is decreasing in P as sum of decreasing functions (15). Moreover, by (15), $F(0) = h/\sigma^2$, and $F(P) = \sum_{r=1}^{R} \gamma_r \sqrt{h\overline{C}_r/P}$ for $P_R \le P$, and (a) follows.

By (11) and (15), we have that $\chi_r = F(P_r)$ for $r = 1, \ldots, R$. This jointly with (a) imply that χ_r is strictly decreasing from $\chi_1 = h/\sigma^2$ to $\chi_{R+1} = 0$, and (b) follows.

By (a) and (b), Eq. (18) has the unique positive root. Moreover, by (b), this root belongs to $[P_{r_*^N}, P_{r_*^N+1}]$. Then, substituting (15) into (18) via straightforward calculation we obtain that this root is given by (19). ■

2.2   Equilibrium Strategies

In this section we derive equilibrium strategies in closed form.

Theorem 1. *In game Γ_N, the jammer's equilibrium strategy $\boldsymbol{J}$ is unique, while user equilibrium strategy is unique except of a particular case (b), where a continuum of user's equilibrium strategies arise. Specifically,*

(a) if

$$h/\sigma^2 < C_P \tag{20}$$

then

$$P = 0 \text{ and } \boldsymbol{J} = \boldsymbol{0} \triangleq \{J_{i,j} = 0 : (i,j) \in \mathcal{N} \times \mathcal{M}\}; \tag{21}$$

(b) if

$$h/\sigma^2 = C_P \qquad (22)$$

then

$$\boldsymbol{J} = \boldsymbol{0} \text{ and } P \text{ is any such such that } P \le P_1; \qquad (23)$$

(c) if

$$h/\sigma^2 > C_P \qquad (24)$$

then $P = P_N$ is uniquely given by (19) and $J_{i,j} = BR_{i,j}(P)$, where

$$J_{i,j} = BR_{J,i,j}(P) = \begin{cases} \frac{1}{g_i}\left(\sqrt{g_i h P / C_{J,j}} - \sigma^2\right), & P > \sigma^4 C_{J,j}/(hg_i), \\ 0, & P \le \sigma^4 C_{J,j}/(hg_i). \end{cases} \qquad (25)$$

Proof: By (2), the payoff of (i,j)-type jammer coincides with jammer's payoff in power control game with complete information [19]. This remark and [19, Lemma 1] imply that the best response of (i,j)-type jammer is given by (25).

Since $v_U(P.\boldsymbol{J})$ is linear in P, P is the best response to a fixed $\boldsymbol{J}$ if and only if the following relation holds:

$$P = \begin{cases} 0, & \sum_{(i,j)\in\mathcal{N}\times\mathcal{M}} \alpha_i\beta_j h/(\sigma^2 + g_i J_{i,j}) < C_P, \\ \text{is any,} & \sum_{(i,j)\in\mathcal{N}\times\mathcal{M}} \alpha_i\beta_j h/(\sigma^2 + g_i J_{i,j}) = C_P, \\ \infty, & \sum_{(i,j)\in\mathcal{N}\times\mathcal{M}} \alpha_i\beta_j h/(\sigma^2 + g_i J_{i,j}) > C_P. \end{cases} \qquad (26)$$

Thus, to find NE we have to solve the best response equations (25) and (26). Two cases arise to consider separately: (I) $P = 0$ and (II) $P > 0$.

(I) Let $P = 0$. Then, by (25), $\boldsymbol{J} = \boldsymbol{0}$. Substituting $P = 0$ and $\boldsymbol{J} = \boldsymbol{0}$ into (26) implies $h/\sigma^2 \le C_P$. Thus, if $h/\sigma^2 < C_P$, then, (a) follows.

(II) Let $P > 0$. Two cases arise to consider: (II-a) $\boldsymbol{J} = \boldsymbol{0}$ and (II-b) $\boldsymbol{J} \ne \boldsymbol{0}$.

(II-a) Let $\boldsymbol{J} = \boldsymbol{0}$. Then, by (26), $h/\sigma^2 = C_P$. Substituting $\boldsymbol{J} = \boldsymbol{0}$ and $P > 0$ into (25) imply that $P \le \sigma^4 C_{J,j}/(hg_i)$ for any i and j. Thus, by (9) and (12), $P \le \sigma^4 \overline{C}_1/h = P_1$, and (b) follows.

(II-b) Let $\boldsymbol{J} \ne \boldsymbol{0}$ and $P > 0$. Then, by (26), $h/\sigma^2 > C_P$. Thus, by (26), we have that $\sum_{(i,j)\in\mathcal{N}\times\mathcal{M}} \alpha_i\beta_j h/(\sigma^2 + g_i J_{i,j}) = C_P$. Substituting (25) into this equation implies (19), and (c) follows from Proposition 1. ∎

3   Stackelberg Game

In this section we consider Stackelberg game (SG) [11] scenario, where the user is leader while each type jammer is follower. Thus, each type jammer implements the best response strategy, and the user, knowing such jammer's behaviour, maximizes its payoff given as follows:

$$\Psi(P) \triangleq v_U(P, \boldsymbol{BR}_J(P)). \qquad (27)$$

Then $(P, \mathbf{BR}_J(P))$ is Stackelberg equilibrium (SE), where $P = \mathrm{argmax}_P \Psi(P)$. Denote this SG by Γ_S.

By (11), (12) and (25), the payoff to the user $\Psi(P)$ can be present as follows:

$$\Psi(P) = \Psi_{r(P)}(P) - C_P P \text{ for } P_{r(P)} \leq P < P_{r(P)+1}, \tag{28}$$

where

$$\Psi_r(P) \triangleq \sum_{i=1}^{r} \gamma_i \sqrt{h \overline{C}_i P} + \frac{hP}{\sigma^2} \sum_{i=r+1}^{R} \gamma_i. \tag{29}$$

3.1   Auxiliary Notations and Results

In this section we introduce auxiliary notations and present auxiliary properties of user's payoff (27).

Proposition 2. *(a) $\Psi(P)$ is continuous for $P \geq 0$ such that $\Psi(0) = 0$ and $\lim_{P \uparrow \infty} \Psi(P) = -\infty$.*
(b) Derivative of $\Psi(P)$ with respect to P is given as follows:

$$\frac{d\Psi}{dP}(P) = \xi(P) - C_P \text{ for } P \in \mathbb{R}_+ \backslash \mathcal{P}, \tag{30}$$

where $\mathcal{P} \triangleq \{P_1, \ldots, P_R\}$ and

$$\xi(P) \triangleq \xi_{r(P)}(P) \text{ for } P_{r(P)} \leq P < P_{r(P)+1} \tag{31}$$

with

$$\xi_r(P) \triangleq \sum_{i=1}^{r} \frac{\gamma_i}{2} \sqrt{\frac{\overline{C}_i h}{P}} + \frac{h}{\sigma^2} \sum_{i=r+1}^{R} \gamma_i \tag{32}$$

and $\xi_{-1}(P) \triangleq \infty$.
(c) $\xi_r(P_r) < \xi_{r-1}(P_r)$.
(d) $\xi(P)$ is strictly decreasing in $\mathbb{R}_+$ and continuous everywhere except on the finite set $\mathcal{P}$.
(e) $\Psi(P) = (h/\sigma^2 - C_P)P$ for $P < P_1$.

Proof: Note that, by (11), we have that $\sqrt{h\overline{C}_r P_r} = \overline{C}_r \sigma^2 = hP_r/\sigma^2$. This and (29) imply that $\Psi(P)$ is continuous. By (29), $\Psi(P) = \sum_{i=1}^{R}(\gamma_i/2)\sqrt{\overline{C}_i h/P} - C_P P$ for $P > P_R$, and (a) follows. (b) follow from (28) and (29).

By (11) and (32), we have that

$$\xi_{r-1}(P_r) - \xi_r(P_r) = \gamma_r \left(\frac{h}{\sigma^2} - \frac{1}{2} \sqrt{\frac{\overline{C}_r h}{P_r}} \right) = \frac{h\gamma_r}{2\sigma^2} > 0,$$

and (c) follows. (d) follows from (30) and (31). (e) follows from (28) and (29). ■

Let us introduce the following auxiliary notations:

$$\overline{\chi}_r \triangleq \xi_r(P_r) = \frac{h}{\sigma^2} \left(\sum_{i=1}^{r} \sqrt{\frac{\overline{C}_i}{\overline{C}_r}} \frac{\gamma_i}{2} + \sum_{i=r+1}^{R} \gamma_i \right), \tag{33}$$

$$\underline{\chi}_r \triangleq \xi_r(P_{r+1}) = \frac{h}{\sigma^2} \left(\sum_{i=1}^{r} \sqrt{\frac{\overline{C}_i}{\overline{C}_{r+1}}} \frac{\gamma_i}{2} + \sum_{i=r+1}^{R} \gamma_i \right), \tag{34}$$

where $r = 0, \ldots, R$. Also, let $\underline{\chi}_{-1} \triangleq \infty$ and $\underline{\chi}_R \triangleq 0$.

Proposition 3. *There is an unique $r_*^S \in \{0, \ldots, R\}$ such that one of the following two relations holds:*

$$\overline{\chi}_{r_*^S} \leq C_P \leq \underline{\chi}_{r_*^S - 1}, \tag{35}$$

$$\underline{\chi}_{r_*^S} < C_P < \overline{\chi}_{r_*^S}. \tag{36}$$

Proof: The result follows from (12), (33) and (34) and Proposition 2(c) and (d). ■

3.2   Stackelberg Equilibrium Strategies

In the following theorem we prove the uniqueness of SE and also present SE in closed form.

Theorem 2. *In game Γ_S there is SE $(P_S, BR_J(P_S))$. Moreover, jammer equilibrium strategy is unique, while user's equilibrium strategy is unique except of a particular case (a), where a continuum of user's equilibrium strategies arise. Specifically,*

(a) if

$$C_P = h/\sigma^2 \tag{37}$$

then there is a continuum of user's equilibrium strategies. Specifically, any strategy P_S such that $P_S \leq P_1$ is equilibrium strategy, while the jammer's equilibrium strategy is unique and it is $J = 0$;

(b) if (37) does not hold then user's equilibrium strategy is unique and it is given
in closed form as follows:

$$
P_S = \begin{cases}
P_{r_*^S}, & \text{if (35) holds,} \\[3ex]
\dfrac{h}{4}\left(\dfrac{\displaystyle\sum_{i=1}^{r_*^S}\gamma_i\sqrt{\overline{C}_i}}{C_P-(h/\sigma^2)\displaystyle\sum_{i=r_*^S+1}^{R}\gamma_i}\right)^2, & \text{if (36) holds,}
\end{cases}
\tag{38}
$$

where r_*^S is given by Proposition 3.

Proof: First note that, by (27) and Proposition 2, there is at least one SE.

To derive SE two cases arise to consider: (a) $h/\sigma^2 \leq C_P$ and (b) $h/\sigma^2 > C_P$.
   (a) Let $h/\sigma^2 \leq C_P$. By (33) and (34), this condition is equivalent to $\overline{\chi}_0 \leq C_P < \infty = \underline{\chi}_{-1}$, Thus, (35) holds for $r_*^S = 0$. Also, by Proposition 2(f) and (32), $\xi(P) < C_P$ for $P = 0$. Thus, by Proposition 2(d), $\Psi(P)$ gets the maximum the the unique point $P = 0 = P_0$, and (38) holds.
   (a) Let $h/\sigma^2 > C_P$. Then, by Proposition 2, we have that there is the unique P_S such that $\Psi(P)$ is increasing for $P < P_S$ and it is decreasing for $P > P_S$. Thus, such $P = P_S$, is the unique equilibrium strategy. Then, two cases arise to consider: (b-i) (35) holds and (b-ii) (36) holds.
   (b-i) Let (35) hold. Then $P_S = P_{r_*^S}$, and (38) follows.
   (b-ii) Let (36) hold. Then $\xi_{r_*^S}(P)$ is continuous and strictly decreasing in $[P_{r_*^S}, P_{r_*^S+1}]$ from $\xi_{r_*^S}(P_{r_*^S}) > C_P$ to $\xi_{r_*^S}(P_{r_*^S+1}) < C_P$ Thus, equation $\xi(P) = C_P$ has the unique root in $[P_{r_*^S}, P_{r_*^S+1}]$, and straightforward calculation implies that this root is given by (38). ∎

4 Discussion of the Results

Let us illustrate the obtained results by an example of a specific network where $n = m = 2$, $\sigma^2 = 1$, $h = 1$, $(g_1, g_2) = (0.5, 1)$, $C_J = (0.5, 1)$, and $\alpha = \beta = (1/3, 2/3)$. Then $\mathcal{N} \times \mathcal{M} = \cup_{r=1}^3 \mathcal{S}_r$, where $S_1 = \{(2, 1)\}, S_2 = \{(1, 1), (2, 2)\}$ and $S_3 = \{(1, 2)\}$. Figure 1 illustrates that an increase in user's transmission cost leads to a decrease in user's NE and SE strategies, and a decrease in its payoff in SG. In NG, due to linear structure of user's payoff on the user's strategy, its equilibrium payoff is equal to zero. Moreover, user's SE strategy is smaller as compared to its NE strategy, and this also follows from (19) and (38). This leads to that jammer's SE strategy is smaller as compared to its NE strategy (see, (25)). Flat segments in agents SE strategies reflect higher sensitiveness of the the NE strategies to varying the network parameters, as compared to the SE strategies (see, also (19) and (38)).

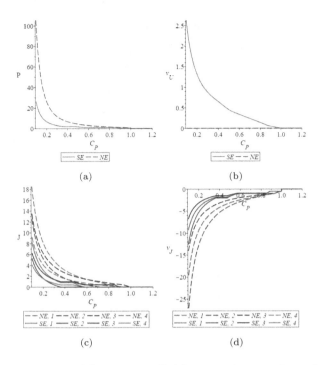

Fig. 1. (a) User's strategy, (b) user's payoff, (c) jammer's strategies and (d) jammer's payoffs.

5   Conclusions

The communication between an user and a receiver in the presence of a jammer, where the jammer, in contrast to the user, has access to local information about jamming fading gain and jamming cost, has been modeled by a Bayesian game in Nash game and Stackelberg game frameworks. The equilibrium are derived in closed form and compared. Higher sensitiveness of the the Nash equilibrium strategies to varying the network parameters, as compared to the Stackelberg equilibrium strategies has been proven and illustrated.

References

1. Al Daoud, A., Alpcan, T., Agarwal, S., Alanyali, M.: A Stackelberg game for pricing uplink power in wide-band cognitive radio networks. In: Proceedings of 47th IEEE Conference on Decision and Control (CDC) (2008)
2. Altman, E., Avrachenkov, K., Garnaev, A.: Jamming in wireless networks under uncertainty. Mob. Netw. Appl. **16**, 246–254 (2011). https://doi.org/10.1007/s11036-010-0272-4
3. Aziz, F., Shamma, J., Stuber, G.L.: Jammer type estimation in LTE with a smart jammer repeated game. IEEE Trans. Veh. Technol. **66**, 7422–7431 (2017)

4. Feng, Z., Ren, G., Chen, J., Zhang, X., Luo, Y., Wang, M., Xu, Y.: Power control in relay-assisted anti-jamming systems: a Bayesian three-layer Stackelberg game approach. IEEE Access **7**, 14623–14636 (2019)

5. Garnaev, A., Petropulu, A., Trappe, W., Poor, H.V.: A power control game with uncertainty on the type of the jammer. In: Proceedings of IEEE Global Conference on Signal and Information Processing (GlobalSIP) (2019)

6. Garnaev, A., Petropulu, A., Trappe, W., Poor, H.V.: A jamming game with rival-type uncertainty. IEEE Trans. Wireless Commun. **19**, 5359–5372 (2020)

7. Garnaev, A., Petropulu, A., Trappe, W., Poor, H.V.: A multi-jammer game with latency as the user's communication utility. IEEE Commun. Lett. **24**, 1899–1903 (2020)

8. Garnaev, A., Petropulu, A., Trappe, W., Poor, H.V.: A switching transmission game with latency as the user's communication utility. In: Proceedings of IEEE International Conference on Acoustics, Speech and Signal Processing (ICASSP) (2020)

9. Garnaev, A., Trappe, W.: The rival might be not smart: revising a CDMA jamming game. In: Proceeding of IEEE Wireless Communications and Networking Conference (WCNC) (2018)

10. Garnaev, A., Trappe, W., Petropulu, A.: Combating jamming in wireless networks: a Bayesian game with jammer's channel uncertainty. In: Proceedings of IEEE International Conference on Acoustics, Speech and Signal Processing (ICASSP), pp. 2447–2451 (2019)

11. Han, Z., Niyato, D., Saad, W., Basar, T., Hjrungnes, A.: Game Theory in Wireless and Communication Networks: Theory, Models, and Applications. Cambridge University Press, New York (2012)

12. He, G., Debbah, M., Altman, E.: k-player Bayesian waterfilling game for fading multiple access channels. In: 3rd IEEE International Workshop on Computational Advances in Multi-Sensor Adaptive Processing (CAMSAP), pp. 17–20 (2009)

13. Jia, L., Yao, F., Sun, Y., Niu, Y., Zhu, Y.: Bayesian Stackelberg game for antijamming transmission with incomplete information. IEEE Commun. Lett. **20**, 1991–1994 (2016)

14. Li, L., Zhang, H., Yang, H., Wan, X.: Security estimation of stochastic complex networks under Stackelberg game framework. In: Proceedings of 37th Chinese Control Conference (CCC) (2018)

15. Sagduyu, Y.E., Berry, R.A., Ephremides, A.: Jamming games in wireless networks with incomplete information. IEEE Commun. Mag. **49**, 112–118 (2011)

16. Scalabrin, M., Vadori, V., Guglielmi, A.V., Badia, L.: A zero-sum jamming game with incomplete position information in wireless scenarios. In: Proceedings of 21th European Wireless Conference, pp. 1–6 (2015)

17. Xiao, L., Chen, T., Liu, J., Dai, H.: Anti-jamming transmission Stackelberg game with observation errors. IEEE Commun. Lett. **19**, 949–952 (2015)

18. Yang, D., Xue, G., Zhang, J., Richa, A., Fang, X.: Coping with a smart jammer in wireless networks: a Stackelberg game approach. IEEE Trans. Wireless Commun. **12**, 4038–4047 (2013)

19. Yang, D., Zhang, J., Fang, X., Richa, A., Xue, G.: Optimal transmission power control in the presence of a smart jammer. In: Proceedings of IEEE Global Communications Conference (GLOBECOM), pp. 5506–5511 (2012)

20. Yuan, L., Wang, K., Miyazaki, T., Guo, S., Wu, M.: Optimal transmission strategy for sensors to defend against eavesdropping and jamming attacks. In: Proceedings of IEEE International Conference on Communications (ICC) (2017)

Demonstrating Spectrally Efficient Asynchronous Coexistence for Machine Type Communication: A Software Defined Radio Approach

Suranga Handagala and Miriam Leeser[✉]

Northeastern University, Boston, MA 02115, USA
handagala.s@ece.neu.edu, mel@coe.neu.edu
https://www.northeastern.edu/rcl/

Abstract. A software defined radio (SDR) approach to demonstrate the coexistence in Machine Type Communication (MTC) scenarios is presented. MTC in recent years has gained significant attention with its inclusion in the 5G business model. Spectrally efficient asynchronous communication is a key enabler in situations involving MTC. Past research has shown that some modifications to baseline cyclic prefix orthogonal frequency division multiplexing (CP-OFDM) can achieve better out-of-band (OOB) suppression and enable asynchronous coexistence. Inspired by this research, we provide a real world example of coexistence using SDR. We demonstrate the ability to asynchronously transmitting waveforms in adjacent channels with very narrow guard bands in between, and still be able to receive and demodulate them with low error vector magnitude (EVM) and low bit error rate (BER) that are comparable to the baseline CP-OFDM that uses synchronous communication.

Keywords: Machine type communication · Software defined radio · FPGA

1 Introduction

The development of 5G has brought new use cases that will support a wide variety of services in the future. The Internet of Things (IoT) for example deviates from the conventional subscriber oriented internet, and uses devices with no human interaction for communication. Designing the physical layer to enable Device to Device (D2D) communication is a challenging task which needs to take into account a number of conflicting trade-offs such as performance, complexity and signaling overhead.

LTE and LTE-Advanced (LTE-A) standards in fourth generation systems have been developed to support increased requirements in capacity and data rates, with their primary target being broadband customers using the mobile internet. LTE is based on the Orthogonal Frequency Division Multiple Access

© ICST Institute for Computer Sciences, Social Informatics and Telecommunications Engineering 2021
Published by Springer Nature Switzerland AG 2021. All Rights Reserved
G. Caso et al. (Eds.): CrownCom 2020, LNICST 374, pp. 69–88, 2021.
https://doi.org/10.1007/978-3-030-73423-7_6

(OFDMA) technique in which multiple users can simultaneously be supported by allocating a subset of subcarriers to each user. OFDMA combined with Multiple Input Multiple Output (MIMO) techniques can provide downlink speeds on the order of hundreds of Mbps which enables mobile broadband services such as video streaming with little or no interruptions. The LTE physical layer is designed in such way that it can dynamically change its modulation type based on user throughput requirements and changing channel conditions by providing adaptive modulation and coding [19]. Despite the well known benefits of LTE, some of its architectural features make it less attractive for MTC applications. In particular, the rectangular pulse shaping used in OFDM makes the physical layer waveform non-spectrally localized causing high OOB radiation due to modulated subcarriers. Although spectrum guardbands can be used to limit such radiation, a loss in spectral efficiency is unavoidable. In millimeter wave system design, spectral efficiency is not a major bottleneck because of the abundance of bandwidth. However, for sub 6 GHz frequencies which are being heavily used in a wide variety of applications, and where even the limited available segments are fragmented, more attention needs be paid as to how the remaining spectrum slices can be utilized in the most efficient manner. Some researchers have proposed alternative waveform design techniques for 5G New Radio (NR) with particular emphasis on IoT applications which require support for multiple asynchronous transmissions. However, despite the fact that OFDM has poor frequency localization, it has been adopted for 5G due to its advantages and backward compatibility reasons. LTE based devices typically use spectral shaping filters in order to meet spectral mask requirements imposed by regulatory agencies. Even so, the occupied bandwidth (OBW) of LTE is limited to 90% of the total bandwidth. Currently, 5G NR has not standardized spectral shaping methods, and as such they are vendor specific. Such flexibility makes it possible for researchers to come up with different spectral shaping methods that may not necessarily be matched between the transmitter and the receiver. During the 5G New Radio (NR) standardization process, several proposals were submitted with modifications to CP-OFDM. Universally Filtered MultiCarrier (UFMC) [15], Filtered OFDM (F-OFDM) [1] and Windowed Overlap and Add (WOLA) OFDM [20] were the main candidates that were considered for the sub 6 GHz spectrum, while proposals such as Filter Bank Multi Carrier (FBMC) have not been considered due to incompatibility with the core OFDM architecture. Extensive studies have been conducted by several researchers to compare merits and demerits of such candidate waveforms in the context of waveform coexistence; i.e., providing support for multiple channels to operate in a time-asynchronous manner without causing significant interference on each other. Most of these studies are based on simulations, while some use over-the-air signals with offline baseband processing using a host PC. To the best of our knowledge there is no platform that supports performing coexistence experiments with real time baseband processing using over-the-air waveforms.

SDR technology has developed significantly over the past few decades due to advancements in analog and digital electronics. Radio frequency integrated circuit (RFIC) technology has enabled integration of multiple discrete components

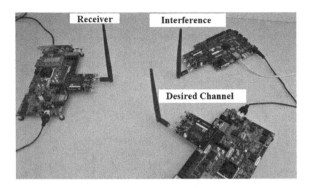

Fig. 1. Interference measurement experimental setup.

in a single chip, and facilitated manufacturing of efficient SDRs with respect to cost and power. In addition, development of Field Programmable Gate Arrays (FPGA), Systems on Chip (SoCs) with associated software tools, and combining them with commercial SDRs have boosted back-end processing capabilities of transmitters and receivers to support implementing modern wireless protocol stacks on such platforms. Case studies relevant to leading SDRs have been presented [17]. While FPGAs in SDR platforms provide processing speeds that are orders of magnitude higher than the maximum data rates of today's wireless standards such as LTE and WiFi, many researchers have not leveraged this capability, and they typically perform back-end processing on a host computer connected to the SDR. The novelty of our work is that we not only use an FPGA for receiver baseband operations, but also provide an interference measurement and quantification setup consisting of three physically distinct SDRs; a transmitter, an interferer and a receiver as shown in Fig. 1. Our contributions are:

- supporting real world LTE based coexistence experiments by using fully separate transmitter/receiver and an interference setup,
- experimentally demonstrating the possibility of coexistence between new and legacy frequency channels,
- supporting higher order QAM constellations by using an accurate channel estimation technique implemented on the FPGA, and
- implementing the F-OFDM filter for receiver-side interference suppression.

Our platform can be used to monitor the performance of a radio link affected by asynchronous interference by using standard performance metrics such as EVM and BER. We investigate such metrics by changing different parameters such as guard band size and interference power. We have also provided some use cases along with experimental results to justify the applicability of our platform in coexistence research.

The rest of the paper is organized as follows. In Sect. 2, we present background related to asynchronous communication as well as related work. We describe the asynchronous coexistence model with mathematical analysis in Sect. 3. Section 4 describes the experimental setup and Sect. 5 presents results. Section 6 concludes and presents future work.

2   Background

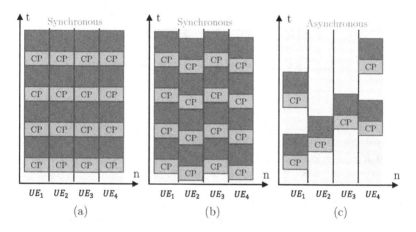

Fig. 2. (a) Theoretically synchronous (b) practically synchronous (c) asynchronous coexistence at the receiver side.

In multiple access situations, base stations (BS) work in a synchronous manner by aligning users' transmit waveforms using a timing advance [8]. LTE BS uses Media Access Control (MAC) layer timing advance control signals to inform the user equipment (UE) when to start transmission in order to maintain orthogonality between transmit waveforms. Consequently, even if two transmit waveforms from two devices are not synchronized at the transmit ends, they arrive synchronously at the receiver. Although synchronous communication is appropriate in a subscriber-based situation, a performance penalty is unavoidable when using it for D2D communication where orders of magnitude more traffic is present. It is expected that the number of IoT devices in the world will be 21.5 billion by 2025. In LTE, a significant amount of control signals are exchanged between the BS and the UE before starting the actual data transfer. This overhead is acceptable in situations involving human-to-human communication in which the connected number of subscribers per BS is not high. In contrast, such overhead in MTC where only sporadic traffic is present, will have detrimental effects on the network performance.

Figure 2 shows how asynchronous multiple access is different from synchronous counterpart when viewed from the BS side when multiple UEs are transmitting. Figure 2(a) shows a theoretical and unrealistic situation where uplink data frames are fully synchronized in a way that no overlapping occurs between UEs. Figure 2(b) shows the practical version of the synchronization where some overlapping is permitted. The amount of overlap may not be higher than some fraction of the CP in order to avoid inter block interference (IBI). Figure 2(c) shows data frames of UEs that are transmitting asynchronously. In OFDM systems, frequency domain equalization is effective only when the interfering signal is synchronous with the desired signal [14]. In asynchronous situations using OFDMA, the receiver side post-equalized symbol EVM increases resulting in high bit errors. Accordingly, synchronization has become a prerequisite to maintain acceptable figures of merit among multiple OFDM channels. However, recent studies have suggested that strict synchronization requirements can be relaxed in OFDMA while still achieving performance comparable to the baseline OFDM [2, 10, 13, 16].

In the context where analytical aspects and simulations related to waveform coexistence have already been studied, and claims on the applicability of some of the waveform candidates have been made, it is important to have a platform to validate such claims at the system level. In what follows, we discuss some SDR platforms which can be used to carry out coexistence related experiments.

2.1 Related Work

AD9361/FMComms3[1] with Xilinx ZC706 is a popular evaluation platform[2] and has been used by several researchers. The authors in [7] have developed a radio virtualization technique using this hardware, which supports multiple standards on the same physical platform. We have used it to demonstrate coexistence of multiple wireless protocols using a single RF front-end [6, 11]. In [21], an experimental testbed using the Universal Software Radio Peripheral (USRP) has been developed to demonstrate the feasibility of multi-user asynchronous access. The authors have characterized different waveform candidates based on their resulting BER and Adjacent Channel Power Ratio (ACPR) taking into account power amplifier nonlinearities. A hardware testbed has been presented in [3] which has been used to conduct asynchronous multiple access experiments using multiple transmitters. It has been shown that OFDM is less robust under lack of synchronization, and waveform candidates such as UFMC and FBMC have shown improved performance in terms of BER. In [18], a generalized fast convolution based filtered OFDM approach for subband filtering in 5G NR has been proposed. In [9], a transparent waveform processing that uses independent transmit and receive processing techniques has been presented which shows that it is possible to have unmatched filtering to improve the BER in asynchronous communication situations.

The work we present in this paper uses the real time receiver baseband processing platform developed by us as the support framework for coexistence

---

[1] https://www.analog.com/en/design-center/evaluation-hardware-and-software/evaluation-boards-kits/eval-ad-fmcomms3-ebz.html.

[2] https://www.xilinx.com/products/boards-and-kits/ek-z7-zc706-g.html.

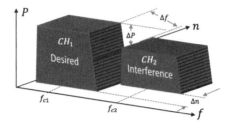

Fig. 3. General asynchronous coexistence model. CH_1 is the desired channel and CH_2 is the interference channel. f, n and P are frequency, discrete time and power, respectively. f_{c1} and f_{c2} are the center frequencies of channels.

experiments [5]. This platform supports over-the-air experiments when natural impairments such as frequency offsets and channel effects are present. In this platform, we have implemented a robust channel estimation technique which helps us demodulate higher order modulated symbols in non-ideal channel conditions. We have added extra functionality on the receiver side of this platform such as filtering in order to improve the BER performance under interference. This platform supports using multiple frequency channels which have software configurable transmit parameters such as center frequency, bandwidth, and consists of a receiver with the ability to perform real time baseband processing on received waveforms. Because of the ability of the receiver to use spectral enhancement filtering methods, we can still leverage the single tap equalization capability of the conventional OFDM to accurately demodulate LTE data channel symbols, and obtain symbols which have satisfactory EVM values even under asynchronous interference.

3   Asynchronous Coexistence

We model the problem of asynchronous coexistence using LTE signals operating in two adjacent channels with configurable relative time offset (Δn), interchannel separation (Δf) and relative power difference (ΔP). This is shown in Fig. 3. In a perfectly time synchronized situation, Δn is equal to 0 which in practical situations is unlikely to happen because of the differences in the multipath structure that different UEs undergo. In general Δn can be either positive or negative. In order to characterize the performance of OFDM for different values of Δn, we use two OFDM sequences; a desired signal and an interference. The duration of the CP is l in both signals. At the receiver, demodulation is performed by extracting the data portion of the desired signal corresponding to the FFT window which typically includes a portion of the CP known as the CP fraction (p). Figure 4 shows the desired and some possible offsets of the same interference signal relative to the desired signal. τ_{max-} and τ_{max+} are maximum negative and positive timing offsets that can be tolerated by the receiver without violating the synchronization requirement. Therefore, in order to achieve synchronous coexistence, it is required that Δn satisfies:

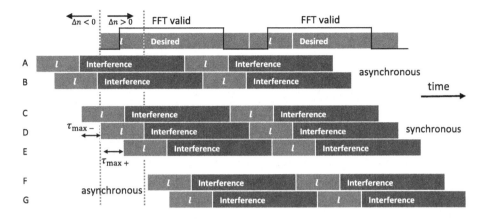

Fig. 4. Illustration of asynchronous and synchronous interference for different values of Δn. The letters A–G show different possible time misalignment situations of the same interference signal with respect to the desired signal. In A and B, $\Delta n < \tau_{max-}$, and in F and G, $\Delta n > \tau_{max+}$ meaning that such situations are asynchronous. C, D and E satisfy Eq. (1), hence are synchronous.

$$\tau_{max-} \leq \Delta n \leq \tau_{max+} \tag{1}$$

where

$$\tau_{max-} = -(1-p)l \tag{2}$$

and

$$\tau_{max+} = pl \tag{3}$$

Figure 5 shows the simulated $EVM_{rms}\%$ vs Δn plot in a situation where CH_1 and CH_2 are present in the medium, and the receiver is configured to the center frequency of CH_1. The power received at the receiver front end from both transmitters are the same. The occupied bandwidth of each channel is 18 MHz and $\Delta f = 0$ kHz. The OFDM blocks that were considered in EVM calculation have $l = 144$ and $p = 0.55$. When $\Delta n < 0$, $\tau_{max-} = -65$ and when $\Delta n > 0$, $\tau_{max+} = 79$. Based on the EVM values, it is clear that asynchronous interference channels increase the receiver side EVM of the desired signal, and this is a bottleneck to coexistence. Filtered OFDM and WOLA OFDM are two main proposals that have been considered to limit OOB radiation and improve the EVM. In F-OFDM, a highly spectally localized spectrum can be achieved using a transmit side low pass filter where the number of coefficients exceed the CP length. It has been shown that such a long filter does not cause significant IBI due to the use of soft truncation time domain windowing. WOLA OFDM is a time domain windowing technique which is used to smooth out discontinuities between successive OFDM blocks by applying a weighted overlap and add (hence the name WOLA) operation. Both these techniques have been extensively studied and their performance in different use cases have been characterized in the literature [4,22].

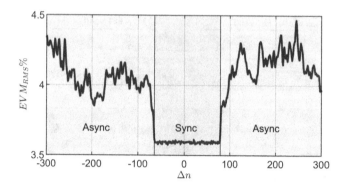

Fig. 5. Plot of percentage EVM vs Δn when $\Delta f = 0\,\text{kHz}$ at an SNR of 30 dB. τ_{max-} and τ_{max+} are 65 and 79, respectively when $p = 0.55$. Note that the EVM is lowest when the interference is synchronous to the desired signal.

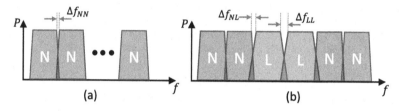

Fig. 6. Coexistence between (a) new (N) and (b) legacy (L) devices. For a target BER specification, $\Delta f_{NN} < \Delta f_{NL} < \Delta f_{LL}$.

By using spectral enhancement, it is expected that new IoT devices will gain the ability to operate with narrower guardbands than legacy devices do. Figure 6 shows two use cases consisting of (a) new users (devices) and (b) combination of new and legacy users. Legacy systems based on OFDM do not have improved frequency localization, thus inter-channel separation, Δf_{LL} needs to be larger than Δf_{NN} to maintain a target BER assuming all other parameters common in both situations remain the same. In contrast, for new devices, Δf_{NN} can be as low as a few subcarriers, but still achieving a comparable BER performance to legacy devices. Another situation considered is the coexistence between new and legacy users. Two situations can be studied here; interference caused by new devices on legacy devices and vice versa. A baseline CP-OFDM pulse is highly time localized which enables low latency communication. The purpose of employing a waveform enhancement technique at the transmitter is to limit the OOB radiation which will affect neighboring channels. In addition, such a technique at the receiver helps to attenuate interference from neighboring channels. Although maintaining orthogonality at the receiver requires using a filter that is matched to the transmitter, in asynchronous communication scenarios, given the fact that orthogonality is already lost, it is not required to enforce a matched filter at the receiver. Such relaxation allows the designer to independently enhance transmit and receiver waveforms by using either filtering or windowing. Figure 7(a)

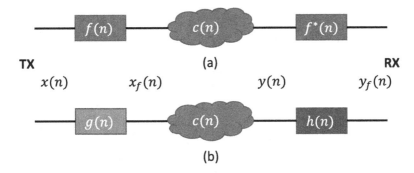

Fig. 7. (a) Matched (b) unmatched waveform processing. $c(n)$ is the channel filter response.

shows the transmit receive setup where the transmit side waveform enhancement technique is matched to that at the receiver. Figure 7(b) shows the unmatched situation in which such techniques do not match. We present in Sect. 5 results obtained using matched and unmatched filtering.

4   SDR Approach for Coexistence Studies

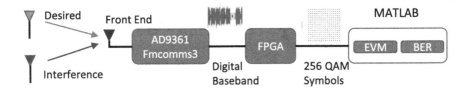

Fig. 8. EVM and BER measurement workflow. The received signal goes through analog and digital signal processing elements of the AD9361, and digital baseband operations such as synchronization and channel estimation are performed on the FPGA in real time.

In our experimental setup, we have one transmitter, one interferer and one receiver as shown in Fig. 1. The receiver front end is an AD9361 FMComms3 SDR connected to Xilinx ZC706 evaluation board which consists of a Zynq 7000 family SoC. The waveforms to be transmitted are generated in MATLAB using LTE Systems Toolbox, and filtered if required. The channel bandwidth is 20 MHz and the center frequency is 3.5 GHz. The distance between the transmitter/interferer and the receiver is 1 m. The transmit power of the interferer relative to the desired signal is changed by writing to attenuation control registers of both devices.

4.1 Workflow

Our workflow is shown in Fig. 8. The desired signal with interference is passed through the analog/digital processing chain of the AD9361. We have implemented baseband processing such as frame detection, synchronization, channel estimation etc. on the FPGA so that the received signals can be processed in real time and symbol constellations can be obtained. The receiver baseband chain has been designed using Simulink and MathWorks HDL Coder. These symbols are processed in MATLAB to calculate EVM and BER. Calculation of the EVM is done by comparing transmitted symbols with received symbols. Usually when symbols are captured at the receiver, they are not aligned with the transmitted symbols. Because of this, we perform a cross correlation between transmitted and received symbols in MATLAB to align them in time. In addition, the received symbols have undergone various levels of scaling due to signal processing taking place on the FPGA. Before calculating the EVM, we need to scale the received symbols to coincide with the transmitted symbols in order to minimize calculation errors.

4.2 F-OFDM Filter Implementation

In this work, we have evaluated both filtering and windowing techniques in simulations and have selected the filtering approach for hardware implementation.

Time Domain Approach: The most straightforward method to implement an F-OFDM filter is by using time domain convolution because of the availability of HDL synthesizable high level blocks such as an HDL optimized discrete FIR filter in Simulink. It is possible in certain situations that implementing time domain convolution is impractical due to limitations of hardware resources available on the chip. In F-OFDM, all filter coefficients are real numbers. Therefore, for a filter with length ($M = 257$), a time domain filter will require 257 DSP48 resources. Allocating this many DSPs could be a challenging task even when modern SoCs are used because there is a possibility of insufficient resources for some of the other operations such as correlation. In such situations, the frequency domain approach may better suit to meet resource constraints of the FPGA.

Frequency Domain Approach: Frequency domain filters take the input signal, perform FFT to transform it into the frequency domain, multiply the signal and the impulse response in the frequency domain and transform back to the time domain using inverse FFT. We use the overlap-save method [12] in which a finite state machine (FSM) generates two overlapping pulses for two FFTs that are used to convert the time domain signal into the frequency domain. The first $M - 1$ valid samples of each IFFT are discarded, and the two outputs are combined to get the filtered signal. The block level implementation of this method is shown in Fig. 9.

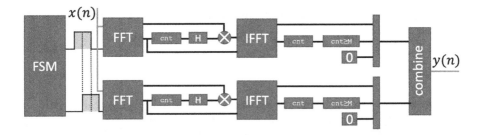

Fig. 9. Frequency domain F-OFDM filter implementation (block level).

We have tested both methods in hardware and verified their operation by achieving comparable EVM performances. The frequency domain method is advantageous in terms of DSP utilization. However, each FFT layer introduces processing latency dependent on the size of the FFT. Therefore, the decision of selecting the most appropriate filtering method needs to be taken after considering these trade-offs.

4.3 An Automated Test Framework for EVM and BER Measurements

The software configurability of the RF front end allows researchers to program AD9361 registers on-the-fly in order to achieve the desired front end behavior. In our experiments, where multiple channels coexist in the medium, it is important to analyze the effect of an interfering signal by changing the attributes of the test setup such as the guard band size, interfering power, etc. Modifying these parameters manually for different use cases is inconvenient and time consuming. Some variables may remain unchanged over multiple experiments while some may change. Such behavior for testing can be achieved in software without much effort by using program control statements such as for loops. We augment the programmability of the SDR by including an additional layer of register programming instructions written for the AD9361 target and run on a host computer that runs MATLAB. Such level of automation allows us to easily increase or decrease the number of experiments covering different use cases, without having to run them manually by configuring parameters for each experiment.

When doing an automated experiment, we preload the waveforms to be transmitted (desired and interference) on the tx side AD9361 targets and control their transmission parameters. ΔP and Δf can be changed by writing to the registers corresponding to the attenuation and center frequency control registers in either the desired or interference transmitter's AD9361. Practically speaking the probability of having a synchronous communication scenario in our test setup is very low. Specifically, this can only happen when the CP of the desired and interference signals overlap subject to the condition given in Eq. (1). The probability of this happening is $144/307200 = 0.0005$ in our case. Therefore, we do not take into account the changes in Δn in our experiments and assume both transmitters to operate asynchronously in practice.

5   Results

In this section we present both simulation and hardware experimental results. We have performed simulations to quantify the effect of synchronous/asynchronous interference in noiseless and noisy situations using different waveform enhancement techniques such as filtering and windowing. We also estimate the probability of bit errors based on the received signal's EVM. We have presented EVM results obtained using over-the-air experiments and real time baseband processing. We show the EVM values obtained with F-OFDM and baseline OFDM, and when legacy channels coexist with legacy and F-OFDM channels at different values of Δf.

5.1   Simulations

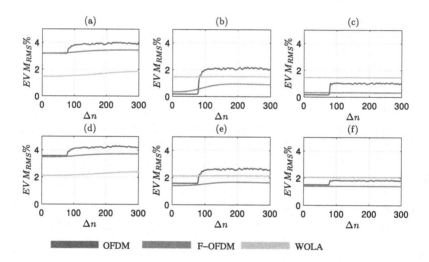

Fig. 10. EVM vs Δn for (a–c) without (d–f) with white noise of SNR $= 30$ dB. Guard band sizes are (a, d) 0, (b, e) 4, and (c, f) 133 subcarriers.

Figure 10(a), (b) and (c) show simulated RMS EVM values of CP-OFDM, F-OFDM and WOLA vs. synchronization error (Δn) obtained for a noiseless channel when guardbands (Δf) are 0, 4, and 133 subcarriers, respectively. F-OFDM filter length M is 257, and the tone offset is 10. WOLA has two overlapping regions with lengths $N_1 = 111$, $N_2 = 112$, with a roll-off factor $\alpha = 0.1$. In the $\Delta f = 0$ situation, the interference from the adjacent channel is expected to be very high because the two channels touch each other in frequency. In LTE with subcarrier spacing equal to 15 kHz, the value 133 translates to a guardband of 2 MHz which is the typical spacing used in practice. In our analysis, $\Delta f = 0$ and 133 can be considered as two extremes with the highest and lowest EVM values, and $\Delta f = 4$ is a situation that gives us an intermediate EVM value. Figure 10(d),(e) and (f) show corresponding results obtained when the SNR is 30 dB. In the noiseless situation, CP-OFDM performs better than F-OFDM and WOLA as long as the orthogonality is maintained between the desired and interfering OFDM blocks; i.e., when Δn is less than pL_{CP} where $p = 0.55$ and $L_{CP} = 144$. However, after the orthogonality is lost when Δn exceeds pL_{CP}, a significant increase in EVM can be noticed. In the case with added noise, F-OFDM outperforms CP-OFDM even in the region where orthogonality is maintained by CP-OFDM. WOLA on the other hand shows high EVM compared to the other two techniques at higher values of Δf and performs better when the spacing is decreased. In practice, it is tolerable to have a small guard band consisting of a few subcarriers; i.e., the spectral efficiency loss incurred is insignificant. Spectral efficiency is a measure of how many bits can be transmitted per second per hertz of bandwidth, and using a few spare subcarriers as a guardband helps improve the EVM without compromising the spectral efficiency too much. In such situations, filtering is favored over windowing, and for hardware experiments using real-time baseband processing, we will only consider filtering instead of windowing. Besides, windowing requires comparatively more changes in the baseline OFDM architecture in order to perform overlap and add operation. In contrast, F-OFDM can be easily integrated as a natural extension to baseline OFDM. For these reasons, we only implement F-OFDM in hardware.

Figure 11a and Fig. 11b show how bit error probability P_b changes with SNR when $\Delta f = 75$ kHz and $\Delta P = 0$ dB for synchronous and asynchronous coexistence scenarios, respectively. These results were obtained using a MATLAB generated baseband LTE signal, which is first modulated to a carrier frequency, and added with an interference signal located at an offset of Δf, then demodulating the affected signal in RF and then in baseband. The character order in the legend represents: transmit side waveform processing technique-interference-receiver side waveform processing technique, respectively. O and F represent non-filtered and filtered techniques, respectively. The black curve represents the baseline BER performance when no interference is present. Note that when synchronous interference is present, OFDM's BER performance curve (O-O-O) closely follows that of the baseline waveform. This is because OFDM maintains orthogonality when properly synchronized which in this case has been achieved. However, in the asynchronous scenario shown in (b), due to the loss of orthogonality, non-filtered OFDM suffers from high BER which is not improved by a significant amount despite the increase in the SNR. In contrast, F-F-F and O-F-F show P_b values on the order of 10^{-6} at an SNR of 30 dB which is comparable to the synchronous situation shown in (a). The performance benefit of the F-F-F situation is obvious; transmit and receive filters are matched and the interference is filtered at the transmit side, causing less OOB radiation into the desired signal

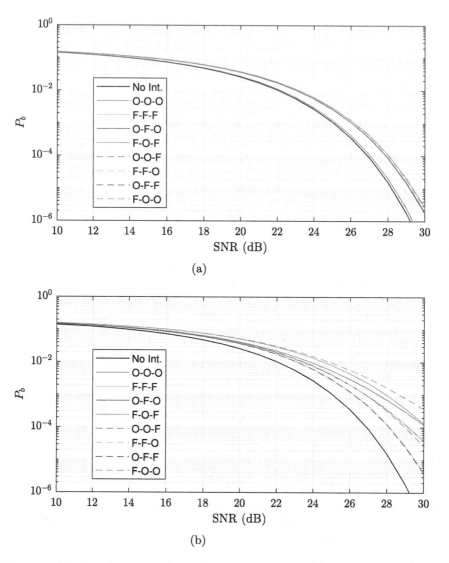

Fig. 11. The effect of (a) synchronous, (b) asynchronous interference on BER performance when $\Delta f = 75\,\text{kHz}$ and $\Delta P = 0\,\text{dB}$.

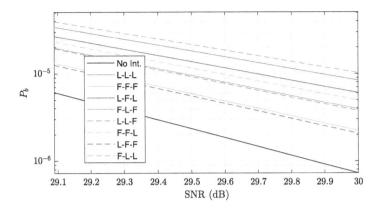

Fig. 12. P_b vs SNR plot for legacy and new channel coexistence scenarios.

band. In O-F-F situation, although the transmit and receive signal processing are unmatched, the BER performance is comparable to that of F-F-F.

Due to the high OOB radiation of OFDM, legacy devices usually employ spectral enhancement techniques such as filters to limit the energy that goes outside their occupied bandwidth. Such filters do not have steeper roll-offs compared to F-OFDM, nevertheless they have improved BER characteristics compared to baseline OFDM with no filtering. We investigate the effect of filtering in legacy OFDM and monitor the BER performance when such channels coexist with legacy and new (F-OFDM) channels. The results are shown in Fig. 12. The legend follows a similar notation to what is shown in Fig. 11a and Fig. 11b; L represents a legacy channel. In the L-L-L situation, P_b at 30 dB of SNR is approximately 8×10^{-6} which is more than a 10 fold improvement compared to the O-O-O situation shown in Fig. 11b. In the L-F-L situation, P_b is further decreased because the radiation of the F-OFDM interferer is further suppressed due to filtering. L-F-F shows the most promising BER performance and is comparable to the baseline non-interference situation. In the L-F-F situation, filtering at the transmitter distorts the constellation less compared to the F-F-F case, hence we notice a slight decrease in P_b compared to that in F-F-F. However the L-F-F situation does not practically exist in existing systems because legacy devices usually use matched filtering at both transmit and receive ends. Nevertheless, such behavior gives us an insight that matched filtering is not a strict requirement to achieve improved BER performance.

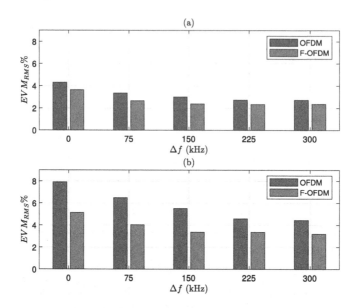

Fig. 13. RMS EVM values obtained after real-time baseband processing: (a) $\Delta P = 0$ dB, (b) $\Delta P = 10$ dB.

5.2 Hardware Experiments

Figure 13(a) and (b) show EVM results for unfiltered and F-OFDM obtained after real time baseband processing when ΔP is 0 and 10 dB, respectively. The waveform processing is matched between the transmitter and the receiver; i.e., for OFDM, we have no filter either at the transmitter or the receiver, and for F-OFDM, transmitter and receiver filters are identical. $\Delta P = 0$ dB means that the received interference power is the same as that of the desired signal, which can occur when two transmitters are equidistant from the receiver and the multipath structure is assumed to be the same for both signals. The $\Delta P = 10$ dB situation can occur when the desired signal path loss is higher than that of the interferer. The highest EVM in both cases occurs at $\Delta f = 0$ kHz because there is no channel spacing and the adjacent channel power leaking into the desired band is significant. When filtering is used, we notice a decrease in EVM of about 15% and 37%, respectively compared to OFDM.

When Δf is increased, EVM decreases as expected. In order to achieve the maximum benefit of F-OFDM, it is required to operate the two channels when Δf is close to 150 kHz, i.e., the bandwidth of 10 LTE subcarriers. In this situation, P_b is approximately 3×10^{-4} and 1×10^{-5} for OFDM and F-OFDM, respectively when $\Delta P = 0$ dB. When Δf is further increased, we get diminishing returns for the F-OFDM filter compared to the baseline OFDM. This is expected because when there is a large guard band, OFDM tends to perform better. Note that selection of Δf is dependent on filter characteristics and BER

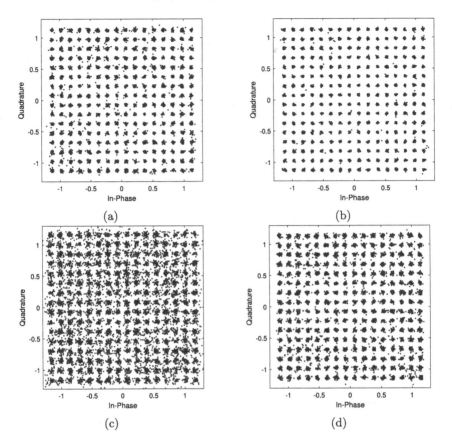

Fig. 14. Receiver side constellations for (a, c) unfiltered (b, d) filtered OFDM when $\Delta f = 150\,\text{kHz}$. ΔP is $0\,\text{dB}$ in (a) and (b), and $10\,\text{dB}$ in (c) and (d).

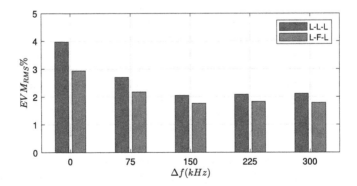

Fig. 15. EVM vs Δf plot for legacy and new channel coexistence scenarios.

requirements, and there is no guarantee that the same value of Δf provides the optimum BER performance in situations which involve different variables.

Figure 14 shows the receiver side constellations obtained after performing FPGA baseband processing on the received signals under different levels of interference when $\Delta f = 150$ kHz. The benefit of filtering is clearly noticeable in (b) and (d) over (a) and (c) in which no filtering is applied. Figure 15 shows EVM values obtained using real time baseband processing for L-L-L and L-F-L situations. According to these results, we notice that because of filtering, new channels induce less interference on legacy channels meaning that new channels can coexist with legacy channels better than legacy channels coexist with each other.

Depending on the nature of the use cases, it is possible to carry out different types of experiments by changing the variables described. The main advantage of using SDR for such experiments is that it provides a real world platform where natural impairments are present, and thus the researcher does not need to mathematically model them. As the technology matures, more capable SDRs with powerful back-end processing devices are expected to come to market. This will provide a great deal of flexibility for wireless communications research.

6   Conclusions and Future Work

We have demonstrated using over-the-air waveforms and real time baseband processing with different use cases that are relevant to demonstrate asynchronous coexistence in the context of machine type communication. The requirement of strict orthogonality in OFDM has been questioned by researchers in the last decade, and it has been pointed out that even under asynchronous communication, it is possible to achieve a target BER performance when appropriate waveform processing techniques are implemented at the transmitter and/or receiver. We have validated such claims by carrying out simulations covering different use cases and have taken a step further by incorporating an SDR testbed to perform experiments involving over-the-air waveforms and real time FPGA based baseband processing. We have presented a system level demonstration covering practical asynchronous communication scenarios in which we comprehensively analyze EVM and BER metrics and shown that it is possible to achieve asynchronous coexistence between different users/devices even when channel separation is only a few kHz.

The platform we presented can be used to carry out different kinds of coexistence related experiments due to the high flexibility of the RF front end. Possible future work is to extend the capabilities of the current setup using the Xilinx RF-SoC[3]. This platform comes with considerably more resources such as DSP48 blocks than the Zynq 7000 family of devices that we currently use. Therefore, it is possible to design and implement longer length filters than what we have already implemented, and achieve even better figures of merit. Another possible

---

[3] https://www.xilinx.com/products/silicon-devices/soc/rfsoc.html.

extension to this research is to demonstrate a real application such as video playback to more clearly highlight the role of interference. For this, we will need to implement the physical downlink shared channel (PDSCH) decoder which will enable playback of the decoded data in real time.

Acknowledgments. This work was supported in part by NSF under Grant CNS-1836880, in part by MathWorks, and by donations from Analog Devices and Xilinx, Inc.

References

1. Abdoli, J., Jia, M., Ma, J.: Filtered OFDM: a new waveform for future wireless systems. In: 2015 IEEE 16th International Workshop on Signal Processing Advances in Wireless Communications (SPAWC), pp. 66–70. IEEE (2015)
2. Bodinier, Q., Bader, F., Palicot, J.: On spectral coexistence of CP-OFDM and FB-MC waveforms in 5G networks. IEEE Access **5**, 13883–13900 (2017)
3. Garcia-Roger, D., de Valgas, J.F., Monserrat, J.F., Cardona, N., Incardona, N.: Hardware testbed for sidelink transmission of 5G waveforms without synchronization. In: 2016 IEEE 27th Annual International Symposium on Personal, Indoor, and Mobile Radio Communications (PIMRC), pp. 1–6 (2016)
4. Guan, P., et al.: 5G field trials: OFDM-based waveforms and mixed numerologies. IEEE J. Sel. Areas Commun. **35**(6), 1234–1243 (2017)
5. Handagala, S., Leeser, M.: Real time receiver baseband processing platform for sub 6 GHz PHY layer experiments. IEEE Access **8**, 105571–105586 (2020)
6. Handagala, S., Mohamed, M., Xu, J., Onabajo, M., Leeser, M.: Detection of different wireless protocols on an FPGA with the same analog/RF front end. In: Moerman, I., Marquez-Barja, J., Shahid, A., Liu, W., Giannoulis, S., Jiao, X. (eds.) CROWNCOM 2018. LNICST, vol. 261, pp. 25–35. Springer, Cham (2019). https://doi.org/10.1007/978-3-030-05490-8_3
7. Jiao, X., Moerman, I., Liu, W., de Figueiredo, F.A.P.: Radio hardware virtualization for coping with dynamic heterogeneous wireless environments. In: Marques, P., Radwan, A., Mumtaz, S., Noguet, D., Rodriguez, J., Gundlach, M. (eds.) CrownCom 2017. LNICST, vol. 228, pp. 287–297. Springer, Cham (2018). https://doi.org/10.1007/978-3-319-76207-4_24
8. Kotzsch, V., Fettweis, G.: Interference analysis in time and frequency asynchronous network MIMO OFDM systems. In: 2010 IEEE Wireless Communication and Networking Conference, pp. 1–6. IEEE (2010)
9. Levanen, T., Pirskanen, J., Pajukoski, K., Renfors, M., Valkama, M.: Transparent Tx and Rx waveform processing for 5G new radio mobile communications. IEEE Wirel. Commun. **26**(1), 128–136 (2018)
10. Medjahdi, Y., et al.: On the road to 5G: comparative study of physical layer in MTC context. IEEE Access **5**, 26556–26581 (2017)
11. Mohamed, M., Handagala, S., Xu, J., Leeser, M., Onabajo, M.: Strategies and demonstration to support multiple wireless protocols with a single RF front-end. IEEE Wirel. Commun. **27**(3), 88–95 (2020)
12. Oppenheim, A.V., Buck, J.R., Schafer, R.W.: Discrete-Time Signal Processing, vol. 2, pp. 558–560. Prentice Hall, Upper Saddle River (2001)
13. Sexton, C., Bodinier, Q., Farhang, A., Marchetti, N., Bader, F., DaSilva, L.A.: Enabling asynchronous machine-type D2D communication using multiple waveforms in 5G. IEEE Internet Things J. **5**(2), 1307–1322 (2018)

14. Thomas, T.A., Vook, F.W.: Asynchronous interference suppression in broadband cyclic-prefix communications. In: 2003 IEEE Wireless Communications and Networking, WCNC 2003, vol. 1, pp. 568–572. IEEE (2003)
15. Vakilian, V., Wild, T., Schaich, F., ten Brink, S., Frigon, J.F.: Universal-filtered multi-carrier technique for wireless systems beyond LTE. In: 2013 IEEE Globecom Workshops (GC Wkshps), pp. 223–228. IEEE (2013)
16. Wunder, G., et al.: 5GNOW: non-orthogonal, asynchronous waveforms for future mobile applications. IEEE Commun. Mag. **52**(2), 97–105 (2014)
17. Wyglinski, A.M., Orofino, D.P., Ettus, M.N., Rondeau, T.W.: Revolutionizing software defined radio: case studies in hardware, software, and education. IEEE Commun. Mag. **54**(1), 68–75 (2016)
18. Yli-Kaakinen, J., Levanen, T., Palin, A., Renfors, M., Valkama, M.: Generalized fast-convolution-based filtered-OFDM: techniques and application to 5G new radio. IEEE Trans. Sig. Process. **68**, 1213–1228 (2020)
19. Yu, C., Xiangming, W., Xinqi, L., Wei, Z.: Research on the modulation and coding scheme in LTE TDD wireless network. In: 2009 International Conference on Industrial Mechatronics and Automation, pp. 468–471. IEEE (2009)
20. Zayani, R., Medjahdi, Y., Shaiek, H., Roviras, D.: WOLA-OFDM: a potential candidate for asynchronous 5G. In: 2016 IEEE Globecom Workshops (GC Wkshps), pp. 1–5. IEEE (2016)
21. Zayani, R., Shaiek, H., Cheng, X., Fu, X., Alexandre, C., Roviras, D.: Experimental testbed of post-OFDM waveforms toward future wireless networks. IEEE Access **6**, 67665–67680 (2018)
22. Zhang, X., Chen, L., Qiu, J., Abdoli, J.: On the waveform for 5G. IEEE Commun. Mag. **54**(11), 74–80 (2016)

Verticals and Applications

Distance Estimation for Database-Assisted Autonomous Platooning

Paweł Kryszkiewicz[(⊠)] , Michał Sybis , Paweł Sroka , and Adrian Kliks

Institute of Radiocommunications, Poznan University of Technology, Poznan, Poland
{pawel.kryszkiewicz,michal.sybis,pawel.sroka,adrian.kliks}@put.poznan.pl

Abstract. This paper presents the improved distance estimation for the purpose of Database-assisted Autonomous Platooning and V2V channel modelling. The proposed approach combines commonly used GPS-based measurements with UWB-based measurements to benefit from both solutions. While GPS allow for unlimited measurement range, the UWB improves accuracy for short range. The paper is based on real-word measurements.

Keywords: Distance estimation · GPS-based measurements · UWB-based measurements · Database assisted platooning · Position and location measurements · Cognitive radio · Vehicular dynamic spectrum access

1 Introduction

Recent years have abounded in many achievements in the field of vehicle traffic automation. Automatic Fare Collection systems, intelligent traffic-light steering or in general Intelligent Transportation Systems (ITS), deployed widely in numerous cities around the globe, are just selected examples of *cities' smartification*. In these cases, the decisions to be made regarding the potential change of the system setup are supported by the infrastructure elements (e.g. road side units mounted on lamps posts or traffic lights), and as such constitute a good example of cooperation between the users and system. However, a significant progress has been made in the last years towards real implementation of autonomous driving concept. Adaptive Cruise Control (ACC) system has been investigated for many years, but recently it has been successively implemented in contemporary cars offered by numerous brands. ACC provides not only the option of speed control (i.e. keeping the value of set velocity or not-exceeding the fixed upper limit). By using dedicated radars or lidars, it controls the distance to the slower car driving ahead. Cooperative version of ACC, known widely as CACC, tends to improve the overall car and system safety level, by means of wireless message exchange between cars within some coverage area. Such an approach is particularly important in case of car platooning, where the *car trains* form a

G. Caso et al. (Eds.): CrownCom 2020, LNICST 374, pp. 91–101, 2021.
https://doi.org/10.1007/978-3-030-73423-7_7

connected convoy, whose behaviour is controlled in wireless way. A number of empirical studies have been performed to evaluate the performance of platooning supported by IEEE 802.11p-based wireless communications, e.g. [1,2].

However, one of the main problems of CACC is the reliability of information exchange between platoon vehicles. From the perspective of data exchange between the platoon members, it is realized by means of short-range wireless communications schemes, such as Dedicated Short-Range Communications (DSRC) or cellular networks (Cellular-V2X, C-V2X). It has been shown, however, that the performance of IEEE 802.11p-based CACC may be significantly reduced as a consequence of even moderate increase in road traffic. This is due to the resultant wireless channel congestion [3–6]. One of the prospective solutions to this problem is to use of a dual-band transceiver that can operate simultaneously in two different frequency bands.

In our approach we concentrate on the so-called Vehicular Dynamic Spectrum Access (VDSA) scheme [7,8], where the traffic may be offloaded from the congested channels to the unoccupied or less occupied frequency bands. The decision of changing channel may be made either solely by the platoon leader or may be supported by the information stored in the context database and delivered to the platoon leader in order to make its decision more reliable. Following the concept of cognitive radio, where secondary network uses frequency bands of the licensed network in an opportunistic way, we select the so-called TV White Spaces (TVWS). In this case, the primary stakeholders are Digital Terrestrial Television (DTT) operators, who broadcast their services to remote users. Moreover, we assume that there exist a database subsystem deployed along the streets to support VDSA procedures.

As the selection of TV band has significant advantage (i.e. stable occupation of the particular TV-channels), there is a strong need to propose a reliable channel propagation model applicable to this band and adjusted to the high-speed scenario. In particular, the typical one or two-slope pathloss models have to be tuned to the platooning scheme, where e.g. the transmitter's and the receiver's speed is high, so the surrounding environment is changing fast. Moreover, the impact of cars present on the propagation path has to be included. However, while performing real-time measurements it appeared that the pahtloss modelling is highly sensitive to localization errors of the transmitter and receiver. The simplest approach, where the position of cars (and in consequence inter-car distance) is derived based on Global Positioning Systems (GPS) signal, cannot be applied when the distances between the cars are small (comparable to typical average GPS location errors). On the other side, application of radar- or lidar-based distance systems is accurate but cannot be applied for larger distances. Thus, there is strong need for a reliable model that allows in consequence for accurate pathloss analysis both in short and long distances. Such a model may be also applied by the database subsystem while evaluation of any distance-related parameters needed in VDSA procedure. In our paper we propose such a model, as well as the algorithm for its application for distance measurements.

The rest of the paper is organized as follows. We first discuss the problem of high impact of measurements inaccuracies on the functioning of VDSA scheme and on designing appropriate pathloss model applicable in such database-oriented system. Next, we present performed experiments and propose the combined approach to distance estimation that may be suitable for both database-assisted platooning and V2V channel modelling, we discuss also its application opportunities. Finally, the paper is concluded.

2 Negative Impact of Distance Measurement Errors

The uncertainty of location of the transmitter and receiver may lead to significant failures in VDSA functioning at least twofold; first, it may influence all location or distance related entries in the database subsystems, e.g. database will provide results based on wrong values, associated with other positions. Second, the models applied in the database (e.g., the applied pathloss model) strongly rely on the accuracy of distance measurements during the conducted measurements campaigns.

2.1 Impact on Distance-Related Entries in Databases

As indicated, in our research We concentrate on utilization of unoccupied television channels known as TVWS [9,10]. In particular, we focus on shifting control channel (CCH) traffic from nominal 5.9 GHz band into the TV band with the support of database assisted system. TVWS are of particular interest, as it may be emphasised that both - the location of the DTT transmitters, and the assignment of TV channels to the certain transmission points, are relatively stable in long time-scale. Thus, such information may be stored in dedicated database subsystem, implemented with the aim of supporting VDSA algorithm [11]. As the access to such data may improve the performance of VDSA scheme, the reliability of this approach highly depends on the accuracy of the database entries. The database used for VDSA may be populated with data gathered in the dedicated measurement campaigns, where the DTT signal power can be detected and stored with presumed spatial resolution. Unfortunately, filling the map with appropriate values of presumed level of confidence requires high accuracy of localization information. Location uncertainty has direct negative impact on the quality of entries stored in the database, as such kind of errors will influence the decisions made by the database system in VDSA scheme. However, in VDSA not only location-dependent data is needed, but also information which rely on distances from a certain location. In other words, it is necessary to achieve reliable distance-related information at given location. For example, it may be necessary to estimate the pathloss between the transmitter and receiver (in our case, platoon leader and other platoon members), which will be later used in the VDSA algorithm and selection of the best TV channel. The pathloss observed between two particular points depends on the calculation of distance, and any errors will have negative impact on the derived pathloss value. However, the problem of location uncertainty appears also in this case, and may even accumulate, when the distance is computed based on the location data stored in the database.

2.2 Impact on Pathloss Modelling

Moreover, as already indicated in the introduction, while performing the pathloss measurements, it is highly important to guarantee high accuracy of the location or distance measurements. Wrong distance measurement will influence the distribution of the measured pathloss values leading to wrong pathloss model. Such wrongly adjusted pathloss model will in turn have negative impact on the database functioning as described above. As distances between the cars in the platoon may vary from some meters to kilometers, there is a need to derive the solution for guaranteeing high accuracy of distance measurements (and modelling) in the entire range of distances.

2.3 Combined Distance Measurement Scheme

The immediate and natural solution is to use navigation-base systems, like GPS to derive the distance between two locations. However, during conducted measurements we have observed that of-the-shelf GPS modules are burdened with some measurement error, whose standard deviation may be unacceptable for databases assisted VDSA especially when the distances are low, i.e. GPS-based distance error is constant with distance that results in very high relative error for short distances and low for significantly distanced vehicles. Thus, we have searched for other technical solutions to proceed with measurement of distance-related values. In particular, the ultra wideband (UWB) system has been utilized to detect the true distance. The experiments have shown that such an approach offers high accuracy (in terms of some centimeters), but it cannot be used for distances longer than tens of meters. In consequence, there is a need for dedicated mathematical model that can combine these two approaches. In this paper, we proposed the combined (GPS and UWB-based) approach to distance estimation that may be suitable for both database-assisted platooning and V2V channel modelling. The need for such a hybrid distance estimation model is caused by constant characteristics of error obtained with GPS-based measurement. Observe that typically in VDSA system the pathloss (in dB) is modeled as linearly dependent from logarithm of distance. An influence of such an error on the results is presented in Fig. 1 for 4 different errors (i.e. $+1$ m, $+4$ m, -1 m and -4 m) affecting the reference $y = x$ line (exact distance without an error). It can be clearly seen that the error is the more significant the lower $log_{10}(x)$ value is. As such, UWB distance measurement system, characterized by significantly low variance but range limited to few tens of meters, can fix the main drawback of GPS-based distance measurements. Moreover, the error is presented in this way (not a commonly used squared error) because now it is easier to observe that even small distance error can significantly affect the slope of the pathloss model used.

The purpose of this paper is to present the improved approach to distance measurements based on GPS and UWB systems data fusion that enhance quality of distance measurement, especially for low distances.

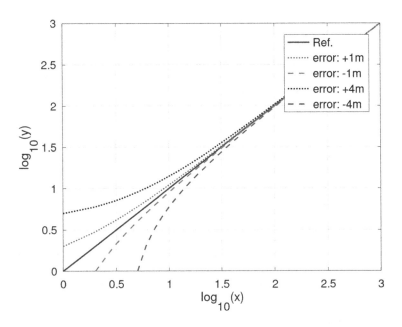

Fig. 1. An exemplary $log_{10}(.)/log_{10}(.)$ plot for constant error introduced.

3 Pathloss Measurements - Conducted Experiment

The channel measurement setup is composed of two cars. The first car (transmitter), was equipped with a USRP N210 module with a WBX extension card and a GPSDO module with an external antenna and a PCTEL LPBMLPVMB/LTE antenna with a gain of 3 dBi. The measured transmitted power is equal to 4.3 dBm, which corresponds to an EIRP value of 7.3 dBm.

The R&S FSL 6 spectrum analyzer connected to a laptop via an Ethernet connection was installed in the second vehicle, acting as a receiver. A wideband AOR 753G antenna with 0 dBi gain was used. In addition, the SYNGIO BU-353 GPS receiver was used. It allowed to determine the location of the vehicle at a given moment in time, as well as its speed and azimuth of movement. Reception processing is based on obtaining time and frequency synchronization according to the note 1MA199, and then calculating the value of the Signal to Noise Ratio (SNR). The receiver sensitivity was established by processing the white noise samples with a scheduler at the input of the spectrum analyzer. In the case of 1000 runs, the estimated SNR values ranged from -17.7 dB to -12.9 dB. It is assumed that the signal is considered correctly received when the SNR value is higher than −13 dB (false alarm probability close to 0.1%). Knowing the transmit power, the total received power and the SNR of the received frame/sequence, the propagation suppression was determined.

In order to allow for improved distance calculation, Ultra Wideband devices, TREK1000 from Decawave [12] were installed. One in each car. The TREK1000 is a wireless transceiver operating according to IEEE802.15.4-2011 standard that allows for range measurement with high accuracy. While one device (in *transmitter* car) was only powered, the second module (in *receiving* car) was connected directly to the laptop for saving logs. As it was observed that the system range is significantly reduced while the UWB antennas are placed inside the cars, about 1.8 m long cables were used to connect antennas placed at the rooftop and UWB transceivers operating inside each car. While the UWB system reports distance about 3.5 times per second, the GPS receivers report positions once every second.

In order to assess correctness of distance measurement between two cars in a dynamic scenario, an experiment was planned. As both distances measured using GPS coordinates (with Vincenty algorithm [13]) and UWB devices can be erroneous, a reference was needed. It was done in a static scenario (called initial experiment) while measuring *real* distance using measuring tape. Three distances between cars were enforced, i.e., 4.5 m, 20.25 m and 44 m. In each case logs from a few minutes were saved for postprocessing. For each distance the offset and standard deviation were estimated (see Table 1). In the last row the table is enriched with mean values.

Table 1. Statistics for GPS and UWB measurements for the initial experiment

	GPS		UWB	
	Offset	Std. dev.	Offset	Std. dev.
Ref. = 4.50 m	2.9855	4.0150	5.0692	0.0284
Ref. = 20.25 m	2.6734	3.8944	4.7326	0.0291
Ref. = 44.00 m	3.1894	5.4421	5.2571	0.0358
Mean value	2.9494	4.3984	5.0196	0.0309

It is observed that for both technologies the calculated offset is fixed, i.e., does not depend from the distance. The offset introduced by the GPS measurements is caused by the fact that in the transmitter car the transmitting antenna (used for V2V channel modeling) was placed at the end of the rooftop while GPS antenna was placed below the car windshield what caused the mismatch in the measurements (due to the use of SUV car the difference is significant). In the case of UWB, offset was introduced by the cables used to connect antennas. Taking into account length of both cables (3.6 m) and average signal propagation velocity (ca. 0.7c) expected offset should be equal to 5.14 m what is almost exactly the value obtained in the experiment. As such the measured offset can be used for data correction. On the other hand, the error measured by standard deviation cannot be easily removed. Based on the above results it can be stated that the distances obtained with the use of UWB are far more precise ($\sigma_{GPS} >> \sigma_{UWB}$) than the results obtained using GPS.

The instantaneous distance measurement error (after removal of the calculated offset) for initial experiment is presented as a function of sample index (time related) in Fig. 2 for the case of 44 m real distance.

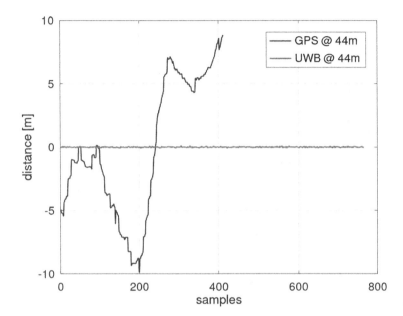

Fig. 2. Fluctuations of the GPS and UWB distance error around the mean value for 44 m static scenario

Obtained curves reflects the values of σ_{GPS} and σ_{UWB} (GPS curve suffer from significant oscillations while UWB have almost no fluctuations). Moreover, it is visible that the GPS-based distance error is significantly correlated in time.

In Fig. 3 a histogram presenting the operation range of both approaches obtained during the main experiment is presented. It needs to be mentioned that the shape of the GPS histogram is affected not by its performance but by the experiment itself. The shape of the UWB histogram clearly shows that useful range is significantly shorter (90% of measurements is below 83.25 m), what highlights the biggest drawback of the used UWB.

Last aspect that also affect the measurements is the number of gaps and its sizes (in the main experiment). Histogram presenting their distribution is shown in Fig. 4.

This figure clearly shows that in the case of GPS-based measurements the gaps in the results are very rare and most of the gaps have a small size. The opposite situation is observed for the UWB-based measurements. In this case very small gaps are very often and wider gaps are also possible.

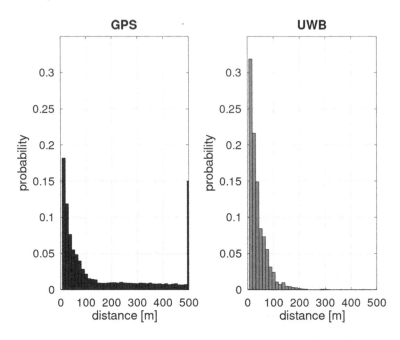

Fig. 3. Histogram of measured distances for GPS (left) and UWB (right) during the main experiment

4   Proposed UWB and GPS-Based Distance Measurements Fusion

As already mentioned, one of the most common approach to measure the distance are the GPS receivers. This approach is relatively simple and offers wide range of measurements, however, is affected with non-sufficient quality. On the other hand UWB based approach implemented in TREK 1000 offers improved measurements quality but significantly lower operating range.

Based on the information presented in previous Section the following algorithm for effective UWB-based and GPS-based combining have been developed.

The first step of the algorithm is to calculate the standard deviation of the UWB-based and GPS-based measurements so these values could be used in the main loop of the algorithm. In the next step a resampling procedure (as a method for signal reconstruction) is performed to handle with short discontinuities. Due to significant error introduced to the output model by this method in the case of insufficient data, this method was limited to 20 samples in the case of UWB measurements (gap length is shorter than ca. 6 s) and 6 samples (corresponds to 6 s) in the case of GPS results. Most suitable approach to find these values is to verify the MSE against the gap size (choose the value just before the increase in MSE become unacceptable). Within the third step of the algorithm the GPS data are up-sampled. The goal of this step is to obtain GPS measurements exactly at the same time instants as for the UWB. Step number four of the algorithm

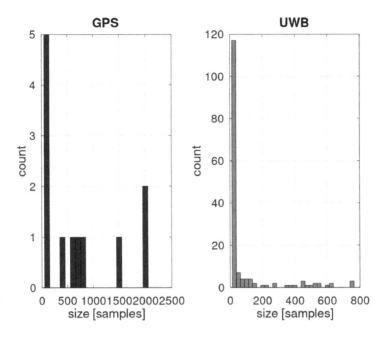

Fig. 4. Histogram of discontinuities for GPS (left) and UWB (right) obtained during the main experiment

combines the data of GPS and UWB. Procedure assumes that if for the given time instant both the measurements are available (GPS and UWB) the output value is calculated with the use of maximum ratio combining with the weights calculated based on the σ_{GPS} and σ_{UWB}. In the case if only GPS or only UWB measurement is available, this measurement is taken directly to the output. If, at this level, neither GPS nor UWB measurement is available algorithm returns no distance for this time instant. Above steps, in general, are able to generate good quality data, however, transition from GPS + UWB to GPS only or from GPS only to GPS + UWB may generate sudden step in output data. To mitigate this disturbance additional filtering with the Hanning window has been performed. The Hanning filter has been feed up with a difference between the GPS and UWB measurements (calculated on last GPS + UWB samples just before the transition to GPS only state or on first samples just after transition from GPS only to GPS + UWB state). Filtered signal, considered as a correcting signal, is subsequently added to the output given after step four. Exemplary curves presenting input GPS and UWB measurements as well as the output data (with and without filtering) for the part of main experiment are shown in Fig. 5.

Based on the above algorithm a statistics showing the percentage contributions of different combinations obtained at the output in the case of main experiment is shown in Table 2.

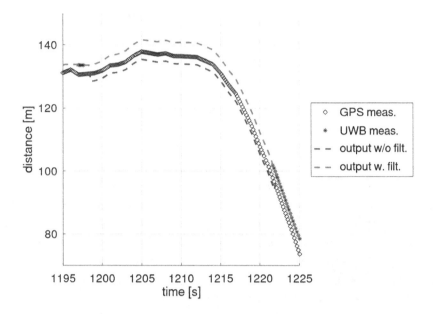

Fig. 5. GPS and UWB input data and calculated distances

Table 2. Different combinations in output distances and its percentage contribution

Type	[%]
GPS + UWB	43.96
GPS only	35.44
UWB only	8.58
No data	12.02

5   Conclusions

In this paper we have proposed a new approach to distance estimation in vehicular applications, such as the database-assisted platooning. It has been shown that the presented solution, relying on the data fusion and analysis from two sources: GPS receiver and UWB-based wireless system, allows to overcome some of the limitations of the individual approaches (i.e. using only GPS or UWB), such as the low accuracy or limited range. The proposed algorithm works in two possible modes: simultaneous use of GPS + UWB or relying on the GPS, with the low-accuracy GPS measurements modified using the estimated correction factor obtained by comparing the results at short distances with ones for the highly-accurate UWB-based system. Furthermore, to ensure smooth transition between the two operation modes, a Hanning window-based filtering has been applied based on a selected number of recent measurements. The described experimental evaluation of the proposed algorithm revealed that it allows to obtain a reliable set of distance measurements both for short and medium distances, thus being an interesting solution for vehicular applications.

Acknowledgement. The work has been realized within the project no. 2018/29/B/ST7/01241 funded by the National Science Centre in Poland and US NSF 1547296.

References

1. Chan, E.: Overview of the sartre platooning project: Technology leadership brief, October 2012
2. Tsugawa, S., Jeschke, S., Shladover, S.E.: A review of truck platooning projects for energy savings. IEEE Trans. Intell. Veh. **1**, 68–77 (2016)
3. Sroka, P., et al.: Szeregowanie transmisji wiadomosci typu BSMw celu poprawy dzialania kooperacyjnego adaptacyjnego tempo-matu. Przeglad Telekomunikacyjny, Wiadomosci Telekomunikacyjne **2017**(6) (2017)
4. Sybis, M., et al.: Communication aspects of a modified cooperative adaptive cruise control algorithm. IEEE Trans. Intell. Transp. Syst. **20**, 1–11 (2019)
5. Rajeswar, R.G., Ramanathan, R.: An empirical study on MAC layer in IEEE 802.11p/WAVE based vehicular ad hoc networks. Procedia Comput. Sci. **143**, 720–727 (2018)
6. Bohm, A., Jonsson, M., Uhlemann, E.: Performance comparison of a platooning application using the IEEE 802.11p MAC on the control channel and a centralized MAC on a service channel. In: 2013 IEEE 9th International Conference on Wireless and Mobile Computing, Networking and Communications (WiMob), pp. 545–552, October 2013
7. Chen, S., Wyglinski, A.M., Pagadarai, S., Vuyyuru, R., Altintas, O.: Feasibility analysis of vehicular dynamic spectrum access via queueing theory model. IEEE Commun. Mag. **49**(11), 156–163 (2011)
8. Chen, S., Vuyyuru, R., Altintas, O., Wyglinski, A.M.: Learning-based channel selection of VDSA networks in shared TV white space. In: 2012 IEEE Vehicular Technology Conference (VTC Fall), September 2012, pp. 1–5 (2012)
9. Harrison, K., Mishra, S.M., Sahai, A.: How much white-space capacity is there? In: 2010 IEEE Symposium on New Frontiers in Dynamic Spectrum (DySPAN), pp. 1–10, April 2010
10. van de Beek, J., Riihijarvi, J., Achtzehn, A., Mahonen, P.: TV whitespace in Europe. IEEE Trans. Mob. Comput. **11**(2), 178–188 (2012)
11. Wei, Z., Zhang, Q., Feng, Z., Li, W., Gulliver, T.A.: On the construction of radio environment maps for cognitive radio networks. In: 2013 IEEE Wireless Communications and Networking Conference (WCNC), pp. 4504–4509, April 2013
12. https://www.decawave.com/
13. Vincenty, T.: Direct and inverse solutions of geodesics on the ellipsoid with application of nested equations. Surv. Rev. **XXIII**(176), 88–93 (1975)

A Priced-Deferred Acceptance (p-DA) Technique for D2D Communication in Factories of the Future

Idayat O. Sanusi[1]([envelope]), Karim M. Nasr[1], and Klaus Moessner[2]

[1] Faculty of Engineering and Science, University of Greenwich, Kent ME4 4TB, UK
{i.o.sanusi,k.m.nasr}@gre.ac.uk
[2] Professorship for Communication Engineering, Faculty of Electrical Engineering and Information Technology, Technical University Chemnitz, Chemnitz, Germany
klaus.moessner@etit.tu-chemnitz.de

Abstract. The Deferred Acceptance (DA) algorithm has often been used to study spectrum sharing and to solve the assignment problem of Device-to-Device (D2D) links co-existing with traditional cellular users. We present a new reward-based DA algorithm denoted as priced-DA or (p-DA) to improve reuse gains in scenarios where there are variations in the number of cellular users and D2D links and when there are variations in the length of the preference lists. Interference coordination is implemented by jointly managing the power allocation and quality of service (QoS) admission control based on the inter-distances between devices. Simulation case studies demonstrate the advantages of the presented p-DA technique in comparison to other tested centralised approaches.

Keywords: Device-to-Device communication · Game theory · Deferred acceptance

1 Introduction

Device-to-Device (D2D) communication is a promising technology aiming to improve spectrum efficiency and to expand network capacity while adapting to the increasing growth of mobile services and data traffic demands. D2D links are usually in close proximity hence enabling increased data rate, lower latency and lower energy consumption [1]. This makes D2D links ideal to support industrial machine-type and vehicular communication applications [2]. D2D links sharing resources in cellular networks are attractive due to the reuse gain and scarcity of spectrum [3]. However, resource-sharing may result in mutual interference between cellular and D2D users. Centralised methods were investigated previously, where the evolved nodeB (eNB) coordinates interference among the users. The centralised schemes often require global acquisition of channel state information (CSI) which often incurs large signalling overheads and high complexity [4]. Many distributed approaches such as game theory which requires partial

© ICST Institute for Computer Sciences, Social Informatics and Telecommunications Engineering 2021
Published by Springer Nature Switzerland AG 2021. All Rights Reserved
G. Caso et al. (Eds.): CrownCom 2020, LNICST 374, pp. 102–111, 2021.
https://doi.org/10.1007/978-3-030-73423-7_8

CSI were also explored. In particular, matching theory has been investigated for allocating cellular resources to D2D users. The deferred acceptance (DA) algorithm was applied to the stable matching problem (a variant of the matching game problem) for resource allocation of D2D users sharing cellular channels and for matching secondary users to primary user channels in cognitive radio networks [5–11]. The deferred acceptance solution to the stable matching problem is typically used for equally-sized opposite sides to be matched. However, the output of the matching may not be optimal in terms of resource sharing if there are variations in sizes of opposite sides and variations in the length of their preference lists.

As an extension to the work presented in [9], this paper studies resource allocation for D2D-links sharing resources with cellular users in a wireless industrial network for factories of the future (FoF). We present a matching game solution denoted as priced-DA (p-DA) which uses a reward-based mechanism to ensure stability and improve resource sharing. The p-DA technique is compared with centralised approaches where no optimisation is implemented in the matching process to reduce complexity and overheads.

The rest of the paper is organised as follows: we present the system model and problem formulation in Sect. 2. We discuss the resource allocation problem in Sect. 3. Examples of case studies and simulation results are presented in Sect. 4. Finally, the main conclusions are presented in Sect. 5.

2 System Model

We consider an industrial factory setting for the uplink of a D2D-enabled cellular network. The network comprises N cellular users (CUEs) denoted by set $C = \{c_1, \ldots, c_i, \ldots c_N\}$ and M D2D users (DUEs) and denoted by set $D = \{d_1, \ldots, d_j, \ldots d_M\}$ deployed with the coverage of the base station. A set of orthogonal sub-channels are available for spectrum allocation and pre-assigned to the CUEs but can be shared with the D2D links once the minimum QoS CUE-DUE resource-sharing partners are guaranteed. We assume spectrum reuse between DUEs and CUEs in uplink frequency division multiplexing (FDD).

The channel gain $g_{c,B}$, between CUE c_i and the base station, B can be expressed as follows:

$$g_{c,B} = C_1 \gamma_{c,B} \chi_{c,B} d_{c,B}^{-\beta_1} \triangleq \zeta_{c,B} d_{c,B}^{-\beta_1} \tag{1}$$

where $\zeta_{c,B} = C_1 \gamma_{c,B} \chi_{c,B}$, C_1 is the pathloss constant determined by system parameters, $\gamma_{c,B}$ is the small-scale fast fading gain due to multipath propagation and assumed to have an exponential distribution with unity mean. The large-scale fading is composed of pathloss with exponent β_1 and shadowing which has slow fading gain $\chi_{c,B}$ with log-normal distribution. $d_{c,B}$ is the distance between CUE c_i and base station B. The channel gain between DUE link d_j of transmitter d_T and receiver d_R is g_{d_T,d_R}, and channel gain of the interference link from d_T to the base station is $g_{d_T,B}$ and from CUE c_i to DUE d_j receiver is g_{c,d_R}.

The SINR at the received at base station B from CUE c_i and at the DUE d_j receiver d_R can be defined as follows:

$$\xi_{c_i} = \frac{P_{c_i} g_{c,B}}{\sigma^2 + \sum_{d_j \in D} \delta_j^i P_{d_j} g_{d_T,B}} \tag{2}$$

$$\xi_{d_j} = \frac{P_{d_j} h_{d_T,d_R}}{\sigma^2 + \sum_{c_i \in C} \delta_j^i P_{c_i} g_{c,d_R}} \tag{3}$$

where P_{c_i} and P_{d_j} are the transmit powers of CUE c_i and DUE d_j respectively, σ^2 is the power of additive white Gaussian noise of each channel. $\delta_j^i \in \{0, 1\}$ is the resource reuse indicator, $\delta_j^i = 1$ if DUE d_j reuses CUE c_i subchannel and is 0 otherwise.

The reliability of DUE d_j is expressed in terms of the maximum tolerable outage probability, p_0. The outage probability constraint, p_R, is given in (4) where P{.} denotes the probability of the input and $\xi_{d_j,\min}$ is the minimum target SINR for d_j.

$$p_R = P\left(\xi_{d_j} \leq \xi_{d_j,\min}\right) \leq p_0 \tag{4}$$

The optimisation objective is to maximise the overall system throughput R. This is formulated as follows using Shannon capacity:

$$\max_{\delta_j^i, P_{c_i}, P_{c_j}} R = B_i \left(\sum_{c_i \in C} \left(\log_2\left(1 + \xi_{c_i}\right) + \sum_{d_j \in D^{\mathbb{A}}} \delta_j^i \log_2\left(1 + \xi_{d_j}\right) \right) \right) \tag{5}$$

subject to

$$\xi_{c_i} \geq \xi_{c_i,\min} \quad \forall c_i \in C \tag{5a}$$

$$\xi_{d_j} \geq \xi_{d_j,\min} \quad \forall d_j \in D^{\mathbb{A}} \tag{5b}$$

$$\delta_j^i p_R \leq p_0 \quad \forall d_j \in D^{\mathbb{A}} \tag{5c}$$

$$P_{c_i} \leq P_{c_i,\max} \quad \forall c_i \in C \tag{5d}$$

$$P_{d_j} \leq P_{d_j,\max} \quad \forall d_j \in D^{\mathbb{A}} \tag{5e}$$

$$\sum_{c_i \in C} \delta_j^i \leq 1 \quad \forall d_j \in D^{\mathbb{A}} \tag{5f}$$

$$\sum_{d_j \in D^{\mathbb{A}}} \delta_j^i \leq 1 \quad \forall c_i \in C \tag{5g}$$

where B_i is the bandwidth, $D^{\mathbb{A}}$ ($D^{\mathbb{A}} \subseteq D$) denotes the set of acceptable DUEs, $\xi_{c_i,\min}$ and $\xi_{d_j,\min}$ is the minimum SINR for c_i and d_j respectively. $P_{c_i,\max}$ and $P_{d_j,\max}$ denote the maximum transmit powers of c_i and d_j respectively.

The minimum QoS of the CUEs and DUEs are given in constraints 5(a)–5(c). The minimum SINR requirements for c_i and d_j respectively are defined 5(a) and 5(b). The reliability requirement for a valid matching between c_i and d_j is defined in 5(c). Constraints 5(d) and 5(e) are to ensure that the transmit of powers of c_i and d_j does not exceed the permitted limits. Constraints 5(f) and 5(g) guarantee a one-to-one pairing between CUEs and DUEs.

The optimisation problem in (5) is a Mixed Integer Non-linear Programming (MINLP) which is NP-hard and cannot solved directly. The solution to the problem is obtained by decomposing it into sub-problems as described in the section that follows.

3   The Resource Allocation Problem

3.1   QoS Admission and Power Allocation

The QoS admission and power allocation is similar to the joint admission and power control (JAPC) presented in [9]. For a CUE c_i to share resources with a DUE d_j, constraints 5(a) to 5(e) must be satisfied. Considering the inter-distances between the devices while relaxing the channel allocation constraints, (6) is obtained as follows.

$$
\begin{cases}
d_{d_T,B} \geq \left(\dfrac{P_{d_j}\xi_{c_i,\min}\zeta_{d_T,B}}{P_{c_i}\zeta_{c,B} - \sigma^2 \xi_{c_i,\min}(d_{c,B})^{\beta_1}} \right)^{\frac{1}{\beta_2}} (d_{c,B})^{\frac{\beta_1}{\beta_2}} \\[3mm]
d_{c,d_R} \geq \left(\dfrac{\xi_{d_j,\min}\left[P_{c_i}\zeta_{c,d_R} + \sigma^2 (d_{c,d_R})^{\beta_3}\right]}{P_{c_i}\zeta_{d_T,d_R}} \right)^{\frac{1}{\beta_4}} (d_{d_T,d_R})^{\frac{\beta_3}{\beta_4}}
\end{cases}
\tag{6}
$$

Setting $\beta_1 = \beta_2$ and $\beta_3 = \beta_4$, (6) implies the distance of the interfering link should be greater than the distance of the intended signal link. The power allocations for which (6) is valid are determined next. The power pair extrema values that can be assigned to c_i and d_j while satisfying (6) are given in 7a, 7b, 7c and 7d.

$$
P_{c_i,\min} = \frac{\xi_{c_i,\min}\left(\sigma^2 + g_{d_T,B}P_{d_j,\max}\right)}{g_{c,B}}
\tag{7a}
$$

$$
P^c_{d_j,\max} = \frac{g_{c,B}P_{c_i,\max} - \sigma^2\xi_{c_i,\min}}{g_{d_T,B}\xi_{c_i,\min}}
\tag{7b}
$$

$$
P_{d_j,\min} = \frac{\xi_{d_j,\min}\left(\sigma^2 + g_{c,d_R}P_{c_i,\max}\right)}{g_{d_T}d_R}
\tag{7c}
$$

$$
P^d_{c_i,\max} = \frac{g_{d_T}d_R P_{d_j,\max} - \sigma^2\xi_{d_j,\min}}{g_{c,d_R}\xi_{d_j,\min}}
\tag{7d}
$$

The set of transmit power extrema for c_i and d_j is represented by P_c and P_d respectively.

$$
P_c = \left\{ P_{c_i,\max}, P^d_{c_i,\max}, P_{c_i,\min} \right\}
\tag{8a}
$$

$$P_d = \left\{ P_{d_j,\max}, P^c_{d_j,\max}, P_{d_j,\min} \right\} \tag{8b}$$

The set P_{cd} denotes the Cartesian product of P_c and P_d which gives the possible set of power pairs for c_i and d_j to share the same sub-channel. The invalid power pairs are eliminated from P_{cd}. A power pair is invalid if any of the transmit powers in the pair exceed maximum transmit power. P^{inv}_{cd} represent the set of invalid power pairs, P^v_{cd} denote the set of valid power pairs.

$$P^v_{cd} = P_{cd} - P^{inv}_{cd} \tag{9}$$

Next, we obtain the set of power pairs, $P^{\xi\min}_{cd}$, for which the minimum SINR threshold values for c_i and d_j are satisfied. The reliability constraint is then evaluated as the outage probability of DUE d_j conditioned on the selected CUE c_i and expressed as [12]. P^R_{cd} $(P^R_{cd} \subseteq P^{\xi\min}_{cd})$ is the set of power pairs for which minimum outage probability of d_j is satisfied. Therefore, CUE c_i and DUE d_j are potential resource-sharing partners if $P^R_{cd} \neq \emptyset$.

The optimal power allocation that maximises the sum throughput subject to the minimum QoS constraints is given as (10)

$$(P^*_{c_i}, P^*_{d_j}) = \arg \max_{(P_{c_i}, P_{d_j}) \in P^R_{cd}} B_i(\log_2(1 + \xi_{c_i}) + \log_2(1 + \xi_{d_j})) \tag{10}$$

Having identified the resource-sharing CUE-DUE pairs that guarantees the minimum QoS requirements and the optimal power assignments to maximise sum rate, the optimal reuse partner for the CUEs $\forall c_i \in C$ is then determined. $S^d_{c_i}$ be the set of acceptable DUEs for CUE c_i and $S^c_{d_j}$ be the set of CUEs that d_j can share resources with. C^A and D^A is the set of all eligible CUEs and acceptable DUEs respectively.

3.2 Priced Deferred Acceptance Game Solution

The resource allocation problem is modeled using the structure of the Stable Marriage Problem (SMP), a one-to-one two-sided matching game approach. The players are a set of eligible CUEs C^A and a set of acceptable DUEs D^A with preference profiles with which they construct their lists of preference partners. The output of the game is the matching of a DUE to a CUE channel.

Utility Function and Preference Profile

The utility that $c_i \in C^A$ generates from sharing its subchannel with DUEs $d_j \in S^d_{c_i}$, is its throughput and given as $u_{c_i}[d_j]$ whereas the utility that any eligible DUE $d_j \in D^A$ obtains from reusing subchannel of CUE $c_i \in S^c_{d_j}$ is its throughput when paired with the CUE and denoted as $u_{d_j}[c_i]$. $\forall c_i \in C^A$ define a strict preference relation $\succ_c$ over a set of DUEs $S^d_{c_i} \subseteq D^A$ such that $d_1 \succ_{c_i} d_2 \Leftrightarrow u_{c_i}[d_1] > u_{c_i}[d_2]$ implies that c_i prefers d_1 to d_2. Similarly, $\forall d_j \in D^A$, a strict preference relation $\succ_d$ is defined over a set of CUEs $S^c_{d_j} \subseteq C^A$ such that $c_1 \succ_{d_j} c_2 \Leftrightarrow u_{d_j}[c_1] > u_{d_j}[c_2]$ implies that d_j prefers c_1 and c_2. $\forall c_i \in C^A$ and $\forall d_j \in D^A$ can construct their preference list $P\ell_{c_i}$ and $P\ell_{d_j}$ by ordering $S^d_{c_i}$ and $S^c_{d_j}$ respectively, giving precedence to ones that provides better utility.

Definition: For $n \neq m$, $\exists c_i \in C^{\mathbb{A}}$ for which $\left| P\ell_{c_i} \right| = 1 = \{d_j\}$ and $c_i \neq \max P\ell_{d_j}$, then $\mu(c_i) = \emptyset$. This implies that if there exist a CUE c_i with only one potential DUE partner d_j in its preference list, and c_i is not the most preferred by d_j then c_i will be unmatched at the output μ. Consequently, the output of the matching may not be optimal in terms of resource-sharing using the traditional DA method as some eligible CUE(s) might not be paired with a suitable partner. To address this challenge and maximise number of eligible CUEs sharing their sub-channels, a 'priced' Deferred Acceptance (p-DA) is presented.

It is assumed that each active UE is charged with a connection fee that corresponds to the achieved data rate. Denote Φ_{c_i} and Φ_{d_j} represent the price charged per connection for the CUEs and DUEs respectively.

$$\begin{cases} \Phi_{c_i} = \pi B_i \log_2(1 + \xi_{c_i}) \\ \Phi_{d_j} = \pi B_i \log_2(1 + \xi_{d_j}) \end{cases} \tag{11}$$

where π is the price per unit rate and assumed to be uniform for all the UEs. Therefore, the total revenue generated by the BS is given by (12).

$$u_B(\Phi) = \sum_{c_i \in C} \Phi_{c_i} + \sum_{d_j \in D^{\mathbb{A}}} \Phi_{d_j} \tag{12}$$

with $1 \leq i \leq N$ and $1 \leq j \leq D_{DA}$, where D_{DA} is the number of admitted DUEs.

To increase the number of CUE-DUE matching and DUE access rate, $d_j \in D^{\mathbb{A}}$ considers the size of the preference list $P\ell_{c_i}$, of $\forall c_i \in C^{\mathbb{A}}$ that proposes at each iteration round and gives precedence to the highest ranked CUE with the least length of $P\ell_{c_i}$. This is because the larger the length of $P\ell_{c_i}$, then $c_i \in C^{\mathbb{A}}$ will have more DUEs to propose to after being rejected in a previous round of proposals and vice versa. At iteration r, $\forall d_j \in D^{\mathbb{A}}$ will consider the proposal of the highest ranked CUE with least length of preference list at iteration $r + 1$, $\left| P\ell_{c_i}^{(r+1)} \right|$, defined in (13)

$$\left| P\ell_{c_i}^{(r+1)} \right| = \left| P\ell_{c_i}^{(r-1)} \right| - d_j^{(r)} \tag{13}$$

The stability of the matching is ensured by using a reward-based mechanism to balance the utility loss of d_j. Since utility is in terms of the achieved rate, d_j will demand from the BS, a reduction in its price which is equivalent to its rate loss from being matched with c_i rather than c_i' else d_j will deviate from the matching.

$R_{d_j}[c_i]$ and $R_{d_j}\left[c_i'\right]$ denote of the achieved rate from $(c_i d_j)$ and $\left(c_i' d_j\right)$ pairing respectively. The rate loss of d_j is given τ and the price of the rate loss is given as follows:

$$\tau = R_{d_j}\left[c_i'\right] - R_{d_j}[c_i] \tag{14}$$

$$\Phi_{\mathcal{L}} = \pi \tau \tag{15}$$

The p-DA algorithm is summarised as follows.

---

p-DA Algorithm

1: Input C^A, D^A, $S_{c_i}^d$ $\forall$ c_i $\in$ C^A, $S_{d_j}^c$ $\forall$ d_j $\in$ D^A and $(P_{c_i}^*, P_{d_j}^*)$ for potential (c_i, d_j) matching.

2: Set up the preference list of eligible CUEs, $P\ell_{c_i}$, by ordering the DUEs with

$$u_{c_i}[d_j] = B_i \log_2 \left(1 + \frac{P_{c_i}^* g_{c,B}}{\sigma^2 + P_{d_j}^* g_{d_T,B}} \right), \ \forall\, c_i \in C^A$$

3: Set up the preference list of acceptable DUEs, $P\ell_{d_j}$, by ordering the CUEs with

$$u_{d_j}[c_i] = B_i \log_2 \left(1 + \frac{P_{d_j}^* g_{d_T,d_R}}{\sigma^2 + P_{c_i}^* g_{c,d_R}} \right), \ \forall\, d_j \in D^A$$

4: Set up a list of unpaired CUEs $U_c = \{c_i : \forall c_i \in C^A\}$

5: **while** $U_c \neq \emptyset$ **do**

6:    c_i proposes and sends $\left| P\ell_{c_i}^{(r+1)} \right|$ to its highest ranked $d_j \in D^A$ that it has not proposed to in its preference list, $\forall\, c_i \in U_c$

7:    **if** $\forall\, d_j \in D^A$ that receives a proposal from $c_i \in U_c$, c_i is the more preferred CUE with the least preference list, $\left| P\ell_{c_i}^{(r+1)} \right|$, compared to its current match c_i'' and c_i' is the most preferred CUE **then**

8:         d_j holds the proposal of c_i and rejects c_i' and c_i'';

9:         $U_c = U_c - c_i$;

10:        $U_c = U_c + c_i'$;

11:        $U_c = U_c + c_i''$;

12:        d_j obtains its rate loss, $\tau = R_{d_j}[c_i'] - R_{d_j}[c_i]$;

13:    **else**

14:        d_j rejects c_i and remain matched to c_i'';

15:    **end if**

16: **end while**

17: output matching μ

---

4   Example Case Studies, Simulation Results and Discussion

The performance of the presented algorithm is verified for an industrial factory setting. The simulation scenario and channel models used is as depicted in [9]. It is assumed that CUEs have been pre-assigned a sub-channel each. The performance of the algorithm is evaluated using the achieved data rates and number of admitted DUEs (or DUE access rate) which is an indication of the number of shared channels. The random approach and the first feasible assignment (FFA) adopt a centralised approach in which the centralised controller matches a DUE to eligible CUE once the minimum QoS criteria are met. No optimisation is considered in the matching. The Random approach matches a CUE to any random DUE while FFA matches a CUE to the first available DUE.

In Fig. 1, we compare the number of admitted DUEs, D_{AD}, for the three algorithms with $N = 50$ and varying M from 10% to 100% of N. D_{AD} remains constant when resource-sharing is not possible due to the violation of QoS requirements as illustrated at $M = 10$ and $M = 20$ for the three algorithms shown. D_{AD} increases as more valid CUE-DUE matchings are established. The p-DA algorithm achieves 12.5% to 14.28% increase in D_{AD} in comparison to the FFA scheme and 28.57% increase compared with random algorithms for $M > 30$ in particular. As the number of DUEs, M, increases, when $D_{AD} = N$, the network gets saturated and no DUE will be able to access a sub-channel as more DUEs are introduced to the system.

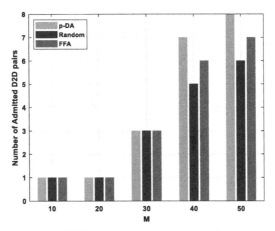

Fig. 1. The number of admitted DUEs, D_{AD} as a function of the number of DUEs, M in the network where $N = 50$, $P_{c_i,\max} = 23$ dBm

The number of admitted DUEs, D_{AD}, directly influences the throughput performance. In Fig. 2, we show the total DUE throughput with respect to M. Random allocation achieves the least performance as expected followed by the FFA algorithm. The performance of p-DA is comparable to the centralised optimisation approach. The p-DA algorithm achieves 9.67% to 14.55% increase in DUE throughput in comparison to the FFA scheme and up to 28.02% random algorithms for $M > 30$ in particular.

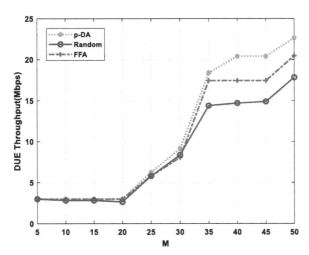

Fig. 2. The total DUE throughput with different number of DUEs, M in the network where $N = 50$, $P_{c_i,\max} = 23$ dBm

In Fig. 3, it is shown that the overall system throughput performance increases with M. The p-DA approach achieves a better performance compared with the random and FFA approaches. The performance of p-DA and FFA algorithms are close however, significant differences are apparent as $M > 35$, where the p-DA scheme shows an improved system throughput.

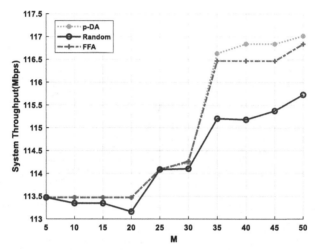

Fig. 3. System throughput with different number of DUEs, M in the network where $N = 50$, $P_{c_i},\text{max} = 23$ dBm

5   Conclusions

In this paper, a priced-deferred acceptance (p-DA) algorithm was presented for a D2D-enabled cellular network targeting wireless industrial scenarios for factories of the future (FoF). The p-DA uses an incentive-based mechanism to ensure stability and improve resource-sharing between cellular and D2D links. Simulations results show that the presented p-DA scheme enhances the D2D throughput performance and reuse gain with lower complexity and signalling overhead compared to the FFA and random allocation schemes. This is because the latter techniques adopt a centralised approach which requires global acquisition of channel information while the p-DA scheme is a distributed approach which relies on local information. Our future work aims at comparing the performance of the presented p-DA game theoretic technique to machine learning approaches.

References

1. Tehrani, M.N., Uysal, M., Yanikomeroglu, H.: Device-to-device communication in 5G cellular networks: challenges, solutions, and future directions. IEEE Commun. Mag. **52**(5), 86–92 (2014)

2. Chen, H., et al.: Ultra-reliable low latency cellular networks: use cases, challenges and approaches. IEEE Commun. Mag. **56**(12), 119–125 (2018)
3. Safdar, G.A., Ur-Rehman, M., Muhammad, M., Imran, M.A., Tafazolli, R.: Interference mitigation in D2D communication underlaying LTE-A network. IEEE Access **4**, 7967–7987 (2016)
4. Li, Z., Guo, C.: Multi-agent deep reinforcement learning based spectrum allocation for D2D underlay communications. IEEE Trans. Vehicular Technol. (2020)
5. Gu, Y., Zhang, Y., Pan, M., Han, Z.: Matching and cheating in device to device communications underlying cellular networks. IEEE J. Sel. Areas Commun. **33**(10), 2156–2166 (2015)
6. Zhang, B., Mao, X., Yu, J.L., Han, Z.: Resource allocation for 5G heterogeneous cloud radio access networks with D2D communication: a matching and coalition approach. IEEE Trans. Vehicular Technol. **67**(7), 5883–5894 (2018)
7. Vassaki, S., Poulakis, M.I., Panagopoulos, A.D.: Spectrum leasing in cognitive radio networks: a matching theory approach. In: Proceedings of 2015 IEEE 81st Vehicular Technology Conference (VTC Spring), pp. 1–5, May 2015
8. Yuan, Y., Yang, T., Xu, Y., Hu, B.: Cooperative spectrum sharing between D2D users and edge-users: a matching theory perspective. In: Proceedings of IEEE 27th Annual International Symposium on Personal, Indoor, and Mobile Radio Communications (PIMRC), pp. 1–6, 4 September 2016
9. Sanusi, I., Nasr, K.M., Moessner, K.: Resource allocation for a reliable D2D enabled cellular network in factories of the future. In: Proceedings IEEE European Conference on Networks and Communications (EuCNC), June 2020
10. Rahim, M., et al.: Efficient Channel Allocation using Matching Theory for QoS Provisioning in Cognitive Radio Networks. Sensors **20**(7), 1872 (2020)
11. Wang, B., Sun, Y., Nguyen, H.M., Duong, T.Q.: A novel socially stable matching model for secure relay selection in D2D communications. IEEE Wirel. Commun. Lett. **9**(2), 162–165 (2020)
12. Liang, L., Li, G.Y., Xu, W.: Resource allocation for D2D-enabled vehicular communications. IEEE Trans. Commun. **65**(7), 3186–3197 (2017)

Data-Driven Intelligent Management of Energy Constrained Autonomous Vehicles in Smart Cities

Yingzhu Ren[1(⊠)], Qimei Cui[1], Xiyu Zhao[1], Yingze Wang[1], Xueqing Huang[2], and Wei Ni[3]

[1] National Engineering Lab for Mobile Network Technologies,
Beijing University of Posts and Telecommunications, Beijing 100876, China
{renyingzhu,cuiqimei}@bupt.edu.cn
[2] New York Institute of Technology, Old Westbury, NY 11568, USA
xhuang25@nyit.edu
[3] Data61, Commonwealth Scientific and Industrial Research Organisation,
Sydney, Australia
Wei.Ni@data61.csiro.au

Abstract. Intelligent transportation is an important component of future smart cities, and electric autonomous vehicles (EAVs) are envisioned to be the main form of transportation because EAVs can save energy, protect the environment, and improve service efficiency. With limited vehicle-specific energy storage capacity and overall constraint in the smart grid's electric load, we propose a novel intelligent management scheme to jointly schedule the travel and charging activities of the EAV fleet in one geographical area. This scheme not only schedules EAVs to meet the passengers' requests but also explores the matching problem between the energy requirement of EAVs and the deployment of charging piles in smart cities. We minimize the total cruise energy consumption of EAVs under the condition of limited energy supply while guaranteeing the quality-of-service (QoS). Network Calculus (NC) is extended to model the electric traffic flow in this paper. With the real-world electric taxi data in Beijing, simulation results demonstrate that the proposed scheme can achieve substantial energy reduction and remarkable improvements in both the order completion rate and utilization rate of the charging stations.

Keywords: Electric autonomous vehicle (EAV) · Intelligent scheduling · Network calculus (NC) · Energy consumption

1 Introduction

1.1 Motivation

Building smart cities are of great significance for sustainable development and enhancing the city's overall competitiveness, and intelligent transportation system plays a key role during this transition [1,2]. With less carbon emission and

G. Caso et al. (Eds.): CrownCom 2020, LNICST 374, pp. 112–125, 2021.
https://doi.org/10.1007/978-3-030-73423-7_9

unified management, electric autonomous vehicles (EAVs) will replace the traditional gas-powered taxi and become the top transportation choice in future smart cities. However, there are still several challenges for the deployment of EAVs, including limited driving range, long recharging duration, inadequate charging stations, and restricted city electricity supply.

1.2 Related Work and Contributions

Intelligent management of EAVs can improve electricity utilization efficiency and alleviate the above challenges. At present, the researches on dispatching schemes for gas-powered taxis are relatively mature and have been widely used in commercial taxi platforms. Meanwhile, for autonomous vehicles (AVs), academic researchers have explored vehicle management, path planning, and unified dispatch problem. Zhang *et al.* has proposed the scheduling strategy for idle AVs [3]. Rule-based scheduling strategies were designed to divide the service area into multiple sub-areas, and the nearest idle vehicles are scheduled for the passengers. If no AV is idle in passenger's located sub-area, neighboring sub-areas will be searched [4–6]. Fagnant *et al.* used a modified Djikstra algorithm to determine the shortest path between an idle AV and a waiting passenger in the actual road network [7]. Bischoff and Maciejewski [8] simulated a city-wide replacement of private cars using autonomous taxi fleets of various sizes. Simulation results suggest that one AV could replace the demand served by ten conventional driven vehicles in Berlin. In our previous work, network calculus (NC) is adopted to model the traffic flows and the management scheme was designed for AVs to reduce the waiting and travel time for passengers [10,11].

The above literature on management has made an ideal assumption that the energy supply for each AV is unlimited. For the scheduling of EVs, however, the energy storage capacity should be taken into consideration. Tseng *et al.* used the Markov decision process to design the optimal path for electric taxis with energy constraints and maximize the profit of taxi drivers [13]. Besides, it is assumed that the EVs could charge in the nearest charging station at any time. Since the idle EAVs can also be arranged to charge, the location and status of the charging station are also significant for the management of EVs [9]. To avoid energy exhaustion before reaching the destination, Sedano *et al.* proposed a reservation plan of charging services for the electric-powered taxis [12]. To minimize the infrastructure investment, Yang *et al.* presented a data-driven optimization algorithm to allocate chargers for the battery electric vehicle (BEV) taxis throughout a city [14]. However, existing electric vehicle scheduling is affected by driver behaviors and cannot achieve large-scale unified scheduling.

From the above observations, we find that the existing work has not studied the large-scale unified scheduling of EAVs by considering both the constraint of energy storage capacity and the limited number of charging piles in each charging station. Therefore, an EAVs' intelligent management system is proposed to minimize the total cruise energy consumption of EAVs under the condition of limited energy and electric load. This intelligent management system provides a look ahead into the solution for a joint deployment of available EAVs in a more practical way. This can benefit the EAVs company for reducing the energy cost and increasing the utilization of charging piles, which can also lower the fee of passengers potentially, making it more feasible to operate the EAVs service in the market.

The contributions of the paper can be summarized as follows.

- Under the energy constraints, An intelligent management scheme to jointly schedule the travel and charging activities of the EAV fleet is proposed to minimize the total cruise energy consumption and match passengers' requests and limited charging piles.
- With EAV-specific energy storage constraints and an overall limited electric load of the urban power grid, we further propose a constrained vehicle dispatching (CVD) algorithm to solve the joint scheduling problem.
- We evaluate the proposed scheduling scheme with simulations on the collected real Beijing electric taxi dataset and demonstrate its effectiveness.

The rest of this paper is organized as follows. In Sect. 2, the collected electric taxi data are introduced along with the system model of NC on how to solve the supply and demand of EAVs for each region. Section 3 details the designed intelligent scheduling scheme of EAVs and simulations over the real electric taxis data. The results are showed and analyzed in Sect. 4. Finally, the whole paper is concluded in Sect. 5.

2   Electric Taxis Dataset and System Model

In order to facilitate the scheduling and management of EAVs, we divide the entire area into grid regions. The EAVs entering and leaving the grids is regarded as the traffic flow model by NC. Considering the limitations of EAVs energy storage and urban electric load, we obtain the energy consumption of EAVs and the location and number of charging piles to ensure all EAVs timely energy replenishment.

In this part, our captured real-world electric vehicle dataset is first introduced along with the distribution of charging stations in the selected observation area. Then, based on the trip records of EVs, the autonomous traffic flow of EAVs is modeled using the Network Calculus method, and the corresponding supply and demand of EAVs are obtained. Lastly, the energy consumption models are further derived for the subsequent energy-aware EAV scheduling.

2.1   Dataset Description

We have obtained a real-world electric taxi dataset, which contains the driving trajectory and electric taxis' behaviors in Beijing. With GPS positioning, the dataset captures 180 million real-time locations and rechargeable battery status with 10 s interval for 30 days: from June 1, 2018, to June 30, 2018, with details given in Table 1. As shown in Fig. 1, the observation area with active electric taxis is evenly divided into $N = 361$ (19×19) grid regions, each about $1\,km^2$. Without considering the behavior of the EVs inside each grid region, we simulate the entire area as a network topology for scheduling large-scale fleets, where each grid is one point. Based on the charging vehicles' location in our dataset, the number of charging piles in each region is approximated as the total number of EVs with charge state being 1, i.e., parking charge. Figure 2 displays the distribution of the number of charging stations in the indexed observation area.

Table 1. Beijing electric taxi dataset

Feature	Definition	
ID	Unique taxi identification	
Timestamp	Data recording time (second)	
Location	Longitude and Latitude	
Taxi State	0: Vacant,	1: Loaded,
	2: Parking,	3: Not in service,
	4: Charging	
Battery State	Residual electricity level (0–100)	
Charge State	1: Parking charge,	2: Driving charge,
	3: No-charging,	4: Charging finished

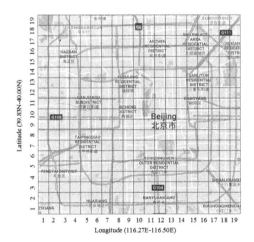

Fig. 1. The observation area with grid regions indexed from left to right and bottom to top, i.e., left bottom corner is region 1, and the top right corner is region 361.

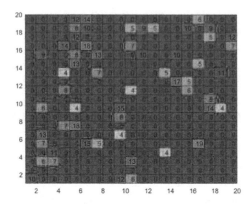

Fig. 2. The number of charging stations in the indexed regions.

2.2   EAV Flow Model

To simulate the computerized autonomous service strategy of EAVs, the NC traffic model in our prior work is extended to model the incoming new requests and outgoing served requests for EAVs [10]. Based on Min-Plus algebra and Max-Plus algebra, NC is an effective method to analyze the performance of a network router by evaluating the incoming and outgoing flows [15].

For a time-slotted system, we denote O_n^t as the optimal number of incoming requests/orders for EAVs in region n during the t-th time slot. O_n^t is borrowed from the effective bandwidth concept in NC, which is used to measure the minimum service rate while keeping the virtual delay of the flow below a predefined threshold [10].

$$O_n^t = \sup_{0 < \tau' \le \tau} (R_n^{t+1} - R_n^t) \times \frac{\tau'}{\tau' + D}, \tag{1}$$

where $\tau = 15\,\text{min}$ is the time duration of one time slot, and $D = 3\,\text{min}$ is the maximum tolerable waiting time for the passengers, we can ensure QoS of EAVs fleet by setting the threshold of D. $(R_n^{t+1} - R_n^t)$ is the number of electric taxis at the n-th region that pick up passengers during time slot t, i.e., the total number of taxis that change state from vacant to loaded.

Let S_n^t denote the number of available EAVs during time slot t in region n. To improve the utilization efficiency of EAVs, we assume that EAVs being charged at charging stations can stop charging at any time and pick up passengers.

$$S_n^t = Q_n^t + C_n^t, \tag{2}$$

where Q_n^t is the total of the number of taxis that change state from loaded to vacant during the time slot t in region n, and C_n^t is the number of charging taxis.

2.3   Energy Models for EAV

We define the set of grid regions as $\mathcal{N} = \{1, \cdots, n, \cdots, N\}$, and the set of time slots as $\mathcal{T} = \{1, \cdots, t, \cdots, T\}$, where T is the total number of time slots. To efficiently schedule the EAVs, the average energy consumed by EAVs to travel from region n_1 to region n_2 is denoted by E_{n_1, n_2}^t, where the energy consumption for two unreachable areas is defined as ∞, and the energy consumption within each region is indicated by 0.

To design an energy-aware EAV scheduling algorithm, in addition to the travel energy model, the battery state of each EAV and the order-specific energy consumption are the other two fundamental inputs, where the first one is given in the original dataset (Table 1), while the latter can be derived by calculating the differences in battery states when driving from one location to another.

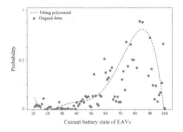
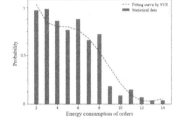

(a) The distribution of vehicles' bat-   (b) The distribution of orders' en-
tery state   ergy consumption $(n = 66)$

Fig. 3. The curve fitted by SVR (09:45 AM–10:00 AM).

The distribution of vehicles' battery state and orders' energy consumption are obtained by curve fitting the Beijing electric taxi dataset, where three weeks of data (75%) as the training set and the remaining one week (25%) for the test dataset. As illustrated in Fig. 3, the fitted curves obtained via support vector regression (SVR) can be used to estimate the EAV' battery state and travel energy consumption in the next time slot [16]. Since the battery state of the EAV cannot change much in a short period, the collected last-minute power distribution of EAVs in the previous time slot can be used as a surrogate for the current time slot of the EAVs. The distribution of the time-varying order energy consumption, however, will be estimated using the time slot-specific fitted curve, as shown in Fig. 3(b).

3   Intelligent Management System

To minimize the total energy consumption of EAVs, an energy-aware scheduling scheme is designed to control the passenger order fulfillment and charging activities of EAVs. The scheme mainly considers the energy constraints of EAVs. First, according to the remaining power of the EAVs, we match the passenger requests and EAVs to minimize the energy consumption of EAVs scheduling while ensuring the completion of the travel. Then, taking into account the urban electric load and the distribution of charging piles to schedule idle EAVs for recharging. The proposed charging scheduling method reduces energy consumption and improves the utilization of charging stations. Since the scheduling scheme will schedule EAVs independently in each time slot, we will drop the time index t in O_n^t, S_n^t, E_{n_1,n_2}^t for the following sections.

3.1   Energy-Aware Passenger Requests Scheduling

The scheduling of EAVs essentially is the matching process between vehicles and passenger requests. When considering the limited driving range and long charging time, it is critical to guarantee that the battery states of all the operating EAVs can last long enough to reach the nearest charging station after dropping off passengers at their destination. Suppose the i-th EAV is scheduled to serve the j-th passenger order, the travel energy consumption from EAV's current region n_i to the passenger's picking up location n_j is E_{n_i,n_j}. After dropping the passenger at the destination n_d, the energy consumption for EAV to reach the nearest charging station's location n_c is E_{n_d,n_c}. The matching process between the vehicles and orders has to guarantee the satisfaction of the following energy constraint.

$$B_i \geq E_j = E_{n_i,n_j} + E_{n_j,n_d} + E_{n_d,n_c} + B_0, \tag{3}$$

where B_i is the battery state of the i-th vehicle and E_j is the energy demand of the j-th order. Both B_i and E_j can be estimated using methods described in Sect. 3.1. B_0 is the lower limit of EAV's battery.

To balance the EAV supply and demand across multiple regions at each time slot, we need to match the optimal number of passenger orders $O = \sum_{n\in\mathcal{N}} O_n$ and the number of available EAVs $S = \sum_{n\in\mathcal{N}} S_n$. The matching problem is modeled as bipartite graph $(\mathcal{V}, \mathcal{P}; \mathcal{E})$.

- $\mathcal{V} = \{1, \cdots, i, \cdots, S\}$ is the set of EAVs available for dispatch at the t-th slot. Each vertex (vehicle) i has two parameters: the current battery state B_i and region n_i.
- $\mathcal{P} = \{1, \cdots, j, \cdots, O\}$ is the set of optimal orders at the t-th slot. Each vertex (order) j also has two parameters: the required energy E_j and the current region n_j.
- $\mathcal{E} = \{E_{n_i,n_j} | i \in \mathcal{V}, j \in \mathcal{P}\}$ is the energy consumed by the i-th EAV to pick up passenger j from location n_j.

At each time slot, the goal of the dispatch algorithm is to determine the minimum-weight matching between EAV $i \in \mathcal{V}$ and passenger order $j \in \mathcal{P}$ that satisfies $B_i \geq E_j$. In our problem setting, this goal can also be interpreted as finding the best action for each EAV to minimize global energy consumption in a coordinated way. We define a binary variable $x_{i,j}$ to indicate whether the vehicle i has been selected to serve the j-th order.

$$
\begin{aligned}
\max_{\{x_{i,j}\}} & \sum_{i \in \mathcal{V}} \sum_{j \in \mathcal{P}} E_{n_i,n_j} x_{i,j} \\
s.t. \ & \sum_{j \in \mathcal{P}} x_{ij} \leq 1, \quad \forall i \in \mathcal{V}, \\
& \sum_{i \in \mathcal{V}} x_{ij} \leq 1, \quad \forall j \in \mathcal{P}, \\
& (B_i - E_j)x_{i,j} \geq 0, \quad \forall i \in \mathcal{V}, \forall j \in \mathcal{P}, \\
& x_{i,j} \in \{0,1\}, \quad \forall i \in \mathcal{V}, \forall j \in \mathcal{P},
\end{aligned}
\tag{4}
$$

where the first constraint specifies each EAV can at most fulfill one customer order, and the second constraint requires that each customer's request can be answered by at most one EAV. The third constraint means for any successfully matched pair, i.e., $x_{i,j} = 1$, the energy gap between the i-th EAV and the j-th order has to be a non-negative number, i.e., the energy supply is no less than the demand.

Without considering the energy gap constraint (the third one) in Eq. (4), the Kuhn-Munkres algorithm can find the pairing between row (EAV) and column (order) over the weight matrix $\mathcal{E}$, such that the total weight (energy consumption) of the selected pairs is minimized [17]. For the constrained minimum-weight matching problem, we have designed the CVD algorithm. In the proposed CVD algorithm, we try to find maximum-weight matching for a bipartite graph with the constraint of energy supply by adding the judgement into the K-M algorithm [18] while dispatching. If the battery state of a vehicle can't satisfy the energy demand of an order, then the vehicle can't be dispatched to the order. Details given in Algorithm 1.

3.2   Grid Load-Aware Charging Scheduling

After the matching process between EAVs and orders described in the previous section, the system is left with a set of EAVs that cannot fulfill the order requirement due to insufficient energy supply. With the location information and battery state for each left-over EAV, a new approach is presented in this section on how to select the optimal charging location from the observation area.

The matching between EAV and charging station is a challenging yet practical problem, because too many EAVs charging at the same time may overload the urban power grid. To improve smart grid performance and economy, the charging schedule of EAVs is constrained by the number and the location of charging stations given in Fig. 2 as well as the capacity of the smart grid [19].

We assume that when the EAV and the matched charging station are in the same region, the energy consumed for the EAV to reach the charging location is negligible. The cross-region charging will occur only when there is no local charging station available. For the cross-region matching between the EAVs and charging stations, it can be treated as the maximum bipartite graph matching problem, similar to the algorithm designed to match EAVs and orders. In particular, we denote E_{n_i,n_c} as the travel energy consumption from the i-th EAV to

---

Algorithm 1. CVD algorithm

Require:

Vehicle set $\mathcal{V}$, order set $\mathcal{P}$, energy matrix $\mathcal{E}$;
Initialize node labels:

$$Label_i = \min_{j \in \mathcal{P}} E_{n_i, n_j}, \forall i \in \mathcal{V}, \text{ and } Label_j = 0, \forall j \in \mathcal{P}$$

Ensure:

The optimal assignment $\mathcal{X} = \{x_{i,j} | i \in \mathcal{V}, j \in \mathcal{P}\}$;
for all $i \in \mathcal{V}$ **do**
2:     Temporary variable $gap \leftarrow \infty$
    for all $j \in \mathcal{P}$ **do**
4:         **if** $Label_i + Label_j - E_{n_i, n_j} = 0$ **and** $B_i \geq E_j$ **then**
            if j has not been served **then**
6:                 Dispatch i to j and update $x_{i,j} = 1$
            else
8:                 Mark j and its dispatched vehicle m
            end if
10:        **else**

$$gap \leftarrow \max\left\{gap, (Label_i + Label_j - E_{n_i, n_j})\right\}$$

12:        **end if**
    end for
14:    **if** i has not been dispatched **then**
        for Marked j **do**
16:            $Label_i \leftarrow Label_i - gap$
            $Label_m \leftarrow Label_m - gap$
18:            $Label_j \leftarrow Label_j + gap$
            Repeat steps 2-23 for vehicle m
            if m could be dispatched to $j' \in \mathcal{P}$, $j' \neq j$ **then**
20:                Update $x_{i,j} = 1$
            end if
22:        **end for**
    end if
24: **end for**

---

the c-th charging pile region. To successfully match the i-th EAV with the c-th charging pile, the battery status of the EAV, i.e., B_i, has to be able to support the selected direct travel route.

$$B_i \geq E_{n_i, n_c}, \tag{5}$$

where without picking up or dropping off the passenger, the requirement on the battery status is less stringent, as compared with Eq. (3).

Similar to Eq. (4), the matching target is to minimize the EAVs' energy consumption. Suppose at each time slot, V_{opt} is the number of routes that are successfully matched by the proposed CVD algorithm, i.e., V_{opt} is the number of EAVs that are scheduled to be charged.

To avoid the overload of urban power grid supply caused by too many electric vehicles being charged at the same time, it is essential to limit the number of charging piles that can operate at the same time. We denote V_{max} as the maximum number of EAVs that can be supported by the grid for simultaneous charging.

$$V_{max} = \frac{L_{Grid}}{L_{EAV}}, \tag{6}$$

where L_{Grid} is the electric load that the smart grid can provide to the charging piles at each time slot, and L_{EAV} is the average charging power of each EAV.

With $V_{max} \geq V_{opt}$, all of the scheduled EAVs can be dispatched directly, following the matching routes. However, if $V_{max} < V_{opt}$, then, V_{max} routes with the least energy consumption will be selected from the scheduling results for dispatch.

4   Simulation Results and Analysis

4.1   The Supply and Optimal Demand of EAVs

Based on the collected electric vehicle dataset in Beijing, we simulate the optimal number of demands for EAVs and the number of served requests for the n-th grid region, $n \in \mathcal{N}$. Figures 4 and 5 show the simulation results of low traffic slot (03:45 AM–04:00 AM) and heavy traffic slot (09:45 AM–10:00 AM), respectively. In addition, battery states will be assigned to the EAVs and the travel energy consumption will be given to the customer orders, based on the fitted curves in Sect. 3.1.

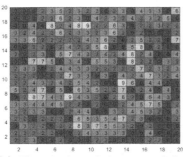

(a) The number of the optimal demand O_n^t

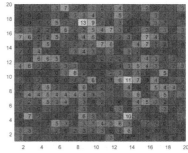

(b) The number of the served requests S_n^t

Fig. 4. Low traffic load: 03:45 AM–04:00 AM, Monday, Nov. 11, 2018.

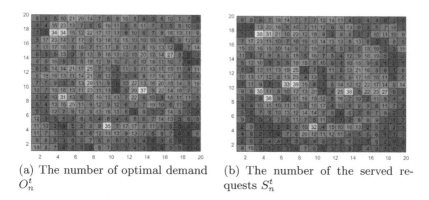

(a) The number of optimal demand O_n^t

(b) The number of the served requests S_n^t

Fig. 5. High traffic load: 09:45 AM–10:00 AM, Monday, Nov. 11, 2018.

By comparing the simulation results in Figs. 4 and 5, we can notice the supply of electric taxis in Beijing is lagging behind the optimal demand. In particular, the supply is about 88.6% of demand during peak hours. Therefore, launching EAVs in the future smart city market still has a lot of potentials. Furthermore, without intervention, the taxi supply and demand in each region are unbalanced, thus vehicle resources and passengers' requirements cannot be well matched under the given QoS.

4.2 Energy-Aware EAV Scheduling

Based on the collected real-world electric vehicle dataset, we match the EAVs and customer orders for two hours with the proposed CVD algorithm. The energy spent on picking up passengers with and without dispatching are shown in Fig. 6(a) and the results indicate that dispatching can reduce energy consumption by 73.5% on average. During rush hours, Fig. 6(b) shows that the number of orders fulfilled with our allocation scheme is about 52% more than that without dispatching. Simulation results demonstrate that the proposed EAV dispatching scheme can reduce vehicle energy consumption while improving passenger satisfaction, thereby greatly increasing the operating company's revenue.

For charging scheduling, although the electric load available at the charging piles can vary with the dynamic demands on the city's smart grid, for simplicity, we set a constant load $L_{Grid} = 500$ MW, $\forall t \in \mathcal{T}$. Today's fast charging piles in Beijing mainly have 60–90 KW power and the average charging power is therefore set as $L_{EAV} = 75$ KW. The energy consumption of EAVs and the utilization rate of charging piles are shown in Fig. 7. It can be observed that charging dispatch can reduce the energy spent on cruising to the charging piles by 10–20%. Moreover, the utilization rate of charging piles is also significantly improved with the proposed charging scheduling algorithm.

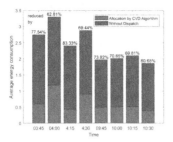   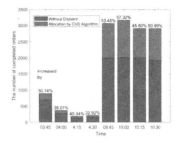

(a) Energy consumption of vehicles   (b) The number of completed orders

Fig. 6. EAV-order matching.

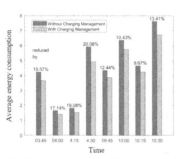   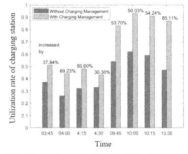

(a) Energy consumption of vehicles   (b) Charging station utilization rate

Fig. 7. EAV-charging station matching.

5   Conclusion

In this paper, we propose an intelligent scheduling system for EAVs while considering the energy-related constraints. The scheduling scheme comprehensively considers the energy-constrained vehicle's order fulfillment and energy replenishment. In the designed framework, EAVs with sufficient energy supply will be matched to fulfill the customer orders, while EAVs with insufficient energy supply will be scheduled to charge at the charging station, which is powered by the urban smart grid. First, the EAVs flow is mathematically modeled using NC, the EAVs' battery states and orders' travel energy consumption are obtained via a machine learning algorithm. To minimize the total energy consumed by EAVs, we propose the CVD algorithm, based on which, the available vehicles and orders are matched with guaranteed QoS, and the charging location and EAVs are matched without violating the overall electric load in the smart grid. Simulation results show that the proposed EAVs' dispatch scheme can save cruising energy and improve both the charging station's utilization rate and the order completion rate.

Acknowledgement. The work was supported by the National Natural Science Foundation of China (61971066) and National Youth Top-notch Talent Support Program.

References

1. Kehua, S., Jie, L., Hongbo, F.: Smart city and the applications. In: International Conference on Electronics, Communications and Control (ICECC), pp. 1028–1031 (2011)
2. Mucahit, K., Haluk, E.: Smart driving in smart city. In: International Istanbul Smart Grid and Cities Congress and Fair (ICSG), pp. 115–119 (2017)
3. Wenwen, Z., Subhrajit, G., Jinqi, F., Ge, Z.: The performance and benefits of a shared autonomous vehicles based dynamic ridesharing system: an agent-based simulation approach. In: Transportation Research Board 94th Annual Meeting (2015)
4. Donna Chen, T., Kockelman, K.M., Hanna, J.P.: Operations of a shared, autonomous, electric vehicle fleet: implications of vehicle & charging infrastructure decisions. Transp. Res. Part A **94**, 243–254 (2016)
5. Daniel Fagnant, J., Kara Kockelman, M.: The travel and environmental implications of shared autonomous vehicles, using agent-based model scenarios. Transp. Res. Part C Emerg. Technol. **40**, 1–13 (2014)
6. Patrick Boesch, M., Francesco, C., Kay, A.W.: Autonomous vehicle fleet sizes required to serve different levels of demand. Transp. Res. Rec. J. Transp. Res. Board **2542**, 111–119 (2016)
7. Fagnant, D.J., Kockelman, K.M.: Dynamic ride-sharing and optimal fleet sizing for a system of shared autonomous vehicles. In: Transportation Research Board 94th Annual Meeting (2015)
8. Joschka, B., Michał, M.: Simulation of city-wide replacement of private cars with autonomous taxis in Berlin. Procedia Comput. Sci. **83**, 237–244 (2016)
9. Han, Y., Zhang, X., Zhang, J., Cui, Q., et al.: Multi-agent reinforcement learning enabling dynamic pricing policy for charging station operators. In: 2019 IEEE Global Communications Conference (GLOBECOM), pp. 1–6 (2019)
10. Qimei, C., Yingze, W., Kwang-Cheng, C., et al.: Big data analytics and network calculus enabling intelligent management of autonomous vehicles in a smart city. IEEE Internet Things J. **6**(2), 2021–2034 (2019)
11. Qimei, C., Ning, W., Martin, H.: Vehicle distributions in large and small cities: spatial models and applications. IEEE Trans. Veh. Technol. **67**(11), 10176–10189 (2018)
12. Sedano, J., Chira, C., Villar, J.R., Ambel, E.M.: An intelligent route management system for electric vehicle charging. Integr. Comput. Aided Eng. **20**(4), 321–333 (2013)
13. Chien-Ming, T., Sid Chi-Kin, C., Xue, L.: Improving viability of electric taxis by taxi service strategy optimization: a big data study of New York city. IEEE Trans. Intell. Transp. Syst. **20**(3), 817–829 (2019)
14. Jie, Y., Jing, D., Liang, H.: A data-driven optimization-based approach for siting and sizing of electric taxi charging stations. Transp. Res. Part C Emerg. Technol. **77**, 462–477 (2017)
15. Le Boudec, J.-Y., Thiran, P.: Network Calculus: A Theory of Deterministic Queuing Systems for the Internet. Springer, Heidelberg (2001). https://doi.org/10.1007/3-540-45318-0

16. Baybulatov, A.A., Promyslov, V.G.: A technique for envelope regression in Network Calculus. In: Application of Information and Communication Technologies (AICT), pp. 1–4 (2017)
17. Haibin, Z., Dongning, L., Siqin, Z., Yu, Z., Luyao, T., Shaohua, T.: Solving the Many to Many assignment problem by improving the Kuhn–Munkres algorithm with backtracking. Theor. Comput. Sci. **618**, 30–41 (2016). https://doi.org/10.1016/j.tcs.2016.01.002
18. James, M.: Algorithms for the assignment and transportation problems. J. Soc. Ind. Appl. Math. **5**(1), 32–38 (1957)
19. Deilami, S., Masoum, A.S., Moses, P.S., Masoum, M.A.: Real-time coordination of plug-in electric vehicle charging in smart grids to minimize power losses and improve voltage profile. IEEE Trans. Smart Grid **2**(3), 456–467 (2011)

A Primer on Large Intelligent Surface (LIS) for Wireless Sensing in an Industrial Setting

Cristian J. Vaca-Rubio[1(✉)], Pablo Ramirez-Espinosa[1], Robin Jess Williams[1],
Kimmo Kansanen[2], Zheng-Hua Tan[1], Elisabeth de Carvalho[1],
and Petar Popovski[1]

[1] Department of Electronic Systems, Aalborg University, Aalborg, Denmark
{cjvr,pres,rjw,zt,edc,petarp}@es.aau.dk
[2] Norwegian University of Science and Technology, Trondheim, Norway
kimmo.kansanen@ntnu.no

Abstract. One of the beyond-5G developments that is often highlighted is the integration of wireless communication and radio sensing. This paper addresses the potential of communication-sensing integration of Large Intelligent Surfaces (LIS) in an exemplary Industry 4.0 scenario. Besides the potential for high throughput and efficient multiplexing of wireless links, an LIS can offer a high-resolution rendering of the propagation environment. This is because, in an indoor setting, it can be placed in proximity to the sensed phenomena, while the high resolution is offered by densely spaced tiny antennas deployed over a large area. By treating an LIS as a radio image of the environment, we develop sensing techniques that leverage the usage of computer vision combined with machine learning. We test these methods for a scenario where we need to detect whether an industrial robot deviates from a predefined route. The results show that the LIS-based sensing offers high precision and has a high application potential in indoor industrial environments.

1 Introduction

Massive multiple-input multiple-output (MIMO) is a fundamental technology in the 5th generation of wireless networks (5G), with the addition of a large number of antennas per base station as its key feature [1]. Looking towards post-5G, researchers are defining a new generation of base stations that are equipped with an even larger number of antennas, giving raise to the concept of large intelligent surface (LIS). Formally, an LIS designates a large continuous electromagnetic surface able to transmit and receive radio waves [2], which can be easily integrated into the propagation environment, e.g., placed on walls. In practice, an

This project has received funding from the European Union's Horizon 2020 research and innovation programme under the Marie Sklodowska-Curie grant agreement No. 813999.

G. Caso et al. (Eds.): CrownCom 2020, LNICST 374, pp. 126–138, 2021.
https://doi.org/10.1007/978-3-030-73423-7_10

LIS is composed of a collection of closely spaced tiny antenna elements. Whilst the performance of LIS in communications has received considerably attention recently [2–5], the potential of these devices could go beyond communications applications, e.g., environment sensing. Indeed, such large surfaces contain many antennas that can be used as sensors of the environment based on the channel state information (CSI).

Sensing strategies based on electromagnetic signals have been thoroughly addressed in the literature in different ways, and applied to a wide range of applications. For instance, in [6], a real-time fall detection system is proposed through the analysis of the communication signals produced by active users, whilst the authors in [7] use Doppler shifts for gesture recognition. Radar-like sensing solutions are also available for user tracking [8] and real-time breath monitoring [9], as well as sensing methods based on radio tomographic images [10,11]. Interestingly, whilst some of these techniques resort solely on the amplitude (equivalently, power) of the receive signals [8,11], in those cases where sensing small scale variations is needed, the full CSI (i.e., amplitude and phase of the impinging signals) is required [9,10].

On a related note, machine learning (ML) based approaches are gaining popularity in the context of massive MIMO systems, providing suitable solutions to optimization problems [12–15]. Due to the even larger dimensions of the system in extra-large arrays, deep learning may play a key role in exploiting complex patterns of information dependency between the transmitted signals.

The popularization of LIS as a natural next step from massive MIMO gives rise to larger arrays and more degrees of freedom, providing huge amounts of data which can feed ML algorithms. Hence, deep learning arises as a potential solution to exploit the performance of LIS.

In this work, we aim to pave the way to the combined use of both deep learning algorithms and the aforementioned large surfaces, exploring, for first time in the literature, the potential of such a joint solution to sense the propagation environment. Specifically, the contribution of this work is twofold:

- We propose an image-based sensing technique based on the received signal power at each antenna element of an LIS. These power samples are processed to generate a high resolution image of the propagation environment that can be used to feed computer vision algorithms to sense large-scale events.
- A computer vision algorithm, based on transfer learning and support vector machine (SVM), is defined to process the radio images generated by the LIS in order to detect anomalies over a predefined robot route.

The performance of the proposed solution is tested in an indoor industrial scenario, where the impact of the array aperture, sampling period and the inter-antenna distance is thoroughly evaluated. We show that both larger apertures and smaller separations between the LIS elements render higher resolution images, improving the performance of the system.

2   Problem Formulation

We consider an industrial scenario where a robot is following a fixed route, and assume that, due to arbitrary reasons, it might deviate from the predefined route and follow an alternative (undesired) trajectory. Hence, our goal is, based on the sensing signal transmitted by the target device, being able to detect whether the robot is following the correct route or not.

In order to perform the anomalous route detection, we assume that an LIS (i.e., a large array of M closely spaced antennas), is placed in the scenario. Therefore, the sensing problem reduces to determine, from the received signal at each of the LIS elements, if the transmission has been made from a point at the desired route, denoted by $\mathbf{p}_c \in \mathbb{R}^3$, or from an anomalous one, denoted by $\mathbf{p}_a \in \mathbb{R}^3$. For the sake of simplicity in a real system implementation, and because we are interested in sensing large scale variations, we resort to the received signal amplitude (equivalently, power). This assumption may lead to simpler system implementations, avoiding the necessity of performing coherent detection.

A classical approach for the aforementioned problem would be performing a hypothesis test based on the received power signal vector. To that end, consider the received complex signal from either $\mathbf{p}_c$ or $\mathbf{p}_a$ to be

$$\mathbf{y}_k = \mathbf{h}_k x + \mathbf{n}_k, \quad k = \{c, a\}, \tag{1}$$

with x the transmitted (sensing) symbol, $\mathbf{h}_k \in \mathbb{C}^{M \times 1}$ the channel vector from each point and $\mathbf{n}_k \sim \mathcal{CN}_M(\mathbf{0}, \sigma^2 \mathbf{I})$ the noise vector. Assume, without loss of generality, that $x = 1$. Hence, the received power vector is given by

$$\mathbf{w}_k = \left(\|y_{1,k}\|^2, \ldots, \|y_{M,k}\|^2 \right)^T, \tag{2}$$

where $y_{i,k}$ for $i = 1, \ldots, M$ are the elements of $\mathbf{y}_k$. The hypothesis test is therefore formulated as

$$\frac{f_{\mathbf{w}_c}(\mathbf{w}|\mathbf{p}_c)}{f_{\mathbf{w}_a}(\mathbf{w}|\mathbf{p}_a)} \underset{\mathbf{p}_a}{\overset{\mathbf{p}_c}{\gtrless}} \frac{P_a}{P_c}, \tag{3}$$

where $f_{\mathbf{w}_k}(\cdot)$ for $k = \{c, a\}$ is the joint probability function of the received signal from each point, $\mathbf{w}$ is the observation vector, and P_a and P_c denote the probability of receiving a signal from $\mathbf{p}_a$ and $\mathbf{p}_c$, respectively. To obtain an optimal estimator, we would need to characterize the joint distribution of the received vector over all the possible anomolaous points, which implies knowing all the possible states of the channels for each path. Also, even in the most simple case, i.e., assuming a pure line-of-sight (LoS) propagation, we would still be unable to distinguish if the two points are in different trajectories or at distinct positions of the same route. Moreover, the a priori probabilities P_a and P_c are needed, which is a non-trivial task.

In a realistic environment, the complexity of the propagation paths is considerable, and the theoretical analysis becomes cumbersome and site-dependent. Hence, in order to gain insight into how the propagation paths between different positions translate into differences in the received signals, we have to resort on machine learning algorithms. This, together with the use of LIS, can provide the necessary information about the propagation environment in order to perform the anomalous route detection.

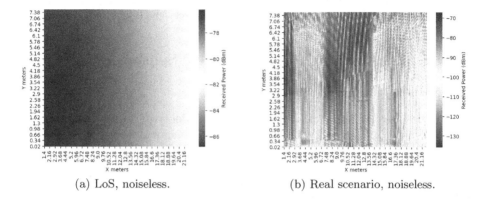

(a) LoS, noiseless.   (b) Real scenario, noiseless.

Fig. 1. Holographic images for LOS and Industry scenarios.

3 Holographic Sensing

A hologram is a recorded interference pattern as a result of constructive and destructive combinations of the superimposed light-wavefronts, i.e., a photographic recording of a light field [16]. In a wireless context, an LIS could be described as a structure which uses electromagnetic signals impinging in a determined scatterer in order to obtain a profile of the environment. That is, we can use the signal power received at each of the multiple elements of the LIS to obtain a high resolution image of the propagation environment. Using this approach, the complexity of the multipath propagation is reduced to using information represented as an image. This provides a twofold benefit: *i)* the massive number of elements that composed the LIS leads to an accurate environment sensing (i.e. high resolution image), and *ii)* it allows the use of computer vision algorithms and image processing techniques to deal with the resulting images.

As an illustrative example, Fig. 1 shows the holographic images obtained from different propagation environments (x and y correspond to the physical dimension of the LIS). Specifically, Figs. 1a correspond to a LoS propagation (no scatterers), whilst Fig. 1b is obtained from an industrial scenario with a rich scattering. Note that, in the case in which different scatterers are placed, their position and shapes are captured by the LIS and represented in the image. To the best of the authors' knowledge, this is the first time that imaged-based sensing is proposed in the literature.

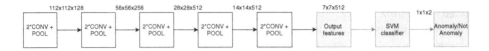

Fig. 2. Proposed model. White and blue blocks refer to VGG19 re-used original architecture and to the additional blocks for our task, respectively. (Color figure online)

4   Machine Learning for Holographic Sensing

4.1   Model Description

We here propose the use of a machine learning model to perform the anomalous route classification task, based on the holographic images obtained at the LIS. In our considered problem, the training data is obtained by sampling the received power at certain temporal instants while the target device is moving along the route. In order to reduce both training time and scanning periods, we resort on transfer learning [17]. Thus, a small dataset can be used, improving the flexibility of the system in real deployments. Among the available strategies for this matter, we will use feature representation.

One of the main requirements for transfer learning is the presence of models that perform well on already defined tasks. These models are usually shared in the form of a large number of parameters/weights the model achieved while being trained to a stable state [18]. The famous deep learning Python library, Keras [19], provides an easy way to reuse some of these popular models. We propose the use of a SVM binary classifier, which has been proved to perform correctly when using a large number of features [20]. In our case, we choose the VGG19 architecture [21].

The model is detailed in Fig. 2. In order to perform the feature extraction, we remove the last fully connected layer (FC) that performs the classification for the purpose of VGG19 and modify it for our specific classification task (anomaly/not anomaly in robot's route). We note that the architecture has been frozen for our case, i.e., the weights and biases in VGG19 are fixed and re-used to generate the features to feed the SVM classifier while the regularization parameter C is tuned to prevent overfitting along the training process.

4.2   Dataset Format

The dataset is obtained by sampling the received signal power at each element of the LIS while the robot moves along the trajectories. Formally, we can define the trajectories as the set of points in the space $\mathbf{P}_t \in \mathbb{R}^{N_p \times 3}$ being N_p the total number of points in the route. Let assume the system is able to obtain N_s samples at each channel coherence interval $\forall\ \mathbf{p}_j \in \mathbf{P}_t$, being $\mathbf{p}_j$ for $j = 1, \ldots, N_p$ an arbitrary point of the route. Hence, the dataset is conformed by $T = N_p \times N_s$ samples (monochromatic holographic image snapshots of received power). Each sample is a gray-scale image which is obtained by mapping the received power into the range of $[0, 255]$. To that end, we apply min-max feature scaling, in which the value of each pixel $m_{i,j}$ for $i = 1, \ldots, M$ and $j = 1, \ldots, N_p$ is obtained as

$$m_{i,j} = \left\lceil m_{\text{MIN}} + \frac{(w_{i,j} - w_{\text{MIN},j})(m_{\text{MAX}} - m_{\text{MIN}})}{w_{\text{MAX},j} - w_{\text{MIN},j}} \right\rceil, \tag{4}$$

where $w_{i,j}$ are the elements of $\mathbf{w}_j$ in (2), i.e. $w_{i,j} = \|h_{i,j} + n_{i,j}\|^2$, $m_{\text{MAX}} = 255$ and $m_{\text{MIN}} = 0$, and

$$w_{\text{MAX},j} = \max_{\{i=1,\ldots,M\}} \mathbf{w}_{i,j}, \quad w_{\text{MIN},j} = \min_{\{i=1,\ldots,M\}} \mathbf{w}_{i,j} \tag{5}$$

are the maximum and minimum received power value from a point $\mathbf{p}_j$ along the surface.

The input structure supported by VGG19 is a RGB image of $n_c = 3$ channels. Due to our monochromatic measurements, our original gray-scale input structure is a one-channel image. To solve this problem, we expand the values by copying them into a $n_c = 3$ channels input structure.

Once the feature extraction is performed, the output is $n_c = 512$ channels of size $n_w = 7$ and $n_h = 7$ pixels. Since SVM works with vectors, the data is reshaped into an input feature vector formed by $7 \times 7 \times 512 = 25088$ features, meaning our dataset is $\{x^{(i)}, y^{(i)}\}_{i=1}^T$, where $x^{(i)}$ is the i-th n-dimensional training input features vector (being $n = 25088$), $x_j^{(i)}$ is the value of the j-th feature, and $y^{(i)}$ is the corresponding desired output label vector.

5   Model Validation

In order to validate the proposed method, we carried out an extensive set of simulations to analyze the performance of the system. To properly obtain the received power values, we use a ray tracing software, therefore capturing the effects of the multipath propagation in a reliable way. Specifically, we consider ALTAIR FEKO WINPROP [22].

5.1   Simulated Scenario

The baseline set-up is described in Fig. 3a, a small size industrial scenario of size $484\,\text{m}^2$. We address the detection of the deviation of the target robot (highlighted in red color) when following a fixed route parallel to the bottom wall, in which the LIS is deployed. The distance between the LIS and the desired trajectory is $13.9\,\text{m}$. For the anomalous routes, a separation of $50/10\,\text{cm}$ have been simulated to analyze the performance of the system when $\Delta d \gg \lambda$ and $\Delta d \approx \lambda$ respectively, as detailed in Fig. 3b.

Table 1. Parameters

Frequency (GHz)	Tx power (dBm)	Nray paths	Antenna type	Antenna spacing (cm)	Propagation model
3.5	20	20	Omni	$\frac{\lambda}{2}/\lambda/2\lambda$	Free space

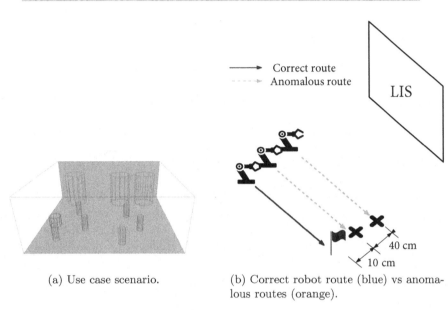

(a) Use case scenario.

(b) Correct robot route (blue) vs anomalous routes (orange).

Fig. 3. Simulated scenario. (Color figure online)

For these routes, we simulate in the ray tracing software N_p points, which corresponds to different positions of the robot in both the correct and anomalous routes. Then, N_s holographic image snapshots of the measurements are taken at every $\mathbf{p}_j$, $j = 1, \ldots, N_p$. The most relevant parameters used for simulation are summarized in Table 1.

In our simulations, we set $N_p = 367$ and $N_s = 10$, thus the dataset is composed of $T = N_p \times N_s = 3670$ radio propagation snapshots containing images of both anomalous and non-anomalous situations, as described in Sect. 4.2. The dataset is split into a 80% training set and 20% for the test set. During the training phase, the obtained optimum regularization value is $C = 0.001$, which was identified by using a 5-fold cross-validation strategy [23].

5.2   Received Power and Noise Modeling

The complex electric field arriving at the i-th antenna element at sample time t, $\widetilde{E}_i(t)$, can be regarded as the superposition of each path, i.e.[1],

$$\widetilde{E}_i(t) = \sum_{n=1}^{N_r} \widetilde{E}_{i,n}(t) = \sum_{n=1}^{N_r} E_{i,n}(t) e^{j\phi_{i,n}(t)}, \tag{6}$$

where N_r is the number of paths and $\widetilde{E}_{i,n}(t)$ is the complex electric field at i-th antenna from n-th path, with amplitude $E_{i,n}(t)$ and phase $\phi_{i,n}(t)$. From (6), and assuming isotropic antennas, the complex signal at the output of the i-th element is therefore given by

$$y_i(t) = \sqrt{\frac{\lambda^2 Z_i}{4\pi Z_0}} \widetilde{E}_i(t) + n_i(t), \tag{7}$$

with λ the wavelength, $Z_0 = 120\pi$ the free space impedance, Z_i the antenna impedance, and $n_i(t)$ is complex Gaussian noise with zero mean and variance σ^2. Note that (7) is exactly the same model than (1); the only difference is that we are explicitly denoting the dependence on the sampling instant t. For simplicity, we consider $Z_i = 1 \, \forall i$. Thus, the power $w_i(t) = \|y_i(t)\|^2$ is used at each temporal instant t to generate the holographic image, as pointed out before. Finally, in order to test the system performance under distinct noise conditions, the average signal-to-noise ratio (SNR) over the whole route, $\overline{\gamma}$, is defined as[2]

$$\overline{\gamma} \triangleq \frac{\lambda^2}{4\pi Z_0 M T \sigma^2} \sum_{t=1}^{T} \sum_{i=1}^{M} |\widetilde{E}_i(t)|^2, \tag{8}$$

where M denotes the number of antenna elements in the LIS.

5.3   Noise Averaging Strategy

Noise is critical in image classification performance [24]. Normally, in the image processing literature, noise removal techniques assume additive noise in the images [25], which is not the case in our system.

Referring to (1) and (7), since we are considering only received powers, the signal at the output of the i-th antenna detector is given by

$$w_i = \left\| \sqrt{\frac{\lambda^2 Z_i}{4\pi Z_0}} \widetilde{E}_i + n_i \right\|^2, \tag{9}$$

---

[1] Note that the electric field also depends on the point $\mathbf{p}_j$. However, for the sake of clarity, we drop the subindex j throughout the following subsections.

[2] This is equivalent to average over all the points $\mathbf{p}_j$ of the trajectory $\mathbf{P}$.

where we have dropped the dependence on t. Also, let assume the system is able to obtain S extra samples at each channel coherence interval $\forall \; \mathbf{p}_j \in \mathbf{P}$. That is, at each point $\mathbf{p}_j$, the system is able to get $N_s' = N_s \times S$ samples. Since the algorithm only expects N_s samples from each point, we can use the extra samples to reduce the noise variance at each pixel. To that end, the value of each pixel $m_{i,j}$ is not computed using directly $w_{i,j}$ as in (4) but instead

$$w_{i,j}' = \frac{1}{S} \sum_{s=1}^{S} w_{i,j,s}, \tag{10}$$

where $w_{i,j,s}$ denote the received signal power at each extra sample $s = 1, \ldots, S$. Note that, if $S \to \infty$, then

$$w_{i,j}'\big|_{S \to \infty} = \mathbb{E}[w_{i,j}|h_{i,j}] = \|h_{i,j}\|^2 + \sigma^2, \tag{11}$$

meaning that the noise variance at the resulting image has vanished, i.e., the received power at each antenna (conditioned on the channel) is no longer a random variable. Observe that the image preserves the pattern with the only addition of an additive constant factor σ^2. This effect is only possible if the system would be able to obtain a very large number S of samples within each channel coherence interval.

5.4   Performance Metrics

To evaluate the prediction effectiveness of our proposed method, we resort on common performance metrics that are widely used in the related literature. Concretely, we are focusing on the F1-Score which is a metric based on the Precision and Recall metrics [26] and is described as:

– Positive F1-Score (PF_1) and Negative F1-Score (NF_1) as the harmonic mean of precision and recall:

$$PF_1 = 2 \cdot \frac{\text{PP} \cdot \text{RP}}{\text{PP} + \text{RP}}, \qquad NF_1 = 2 \cdot \frac{\text{PN} \cdot \text{RN}}{\text{PN} + \text{RN}}. \tag{12}$$

Where PP and RP stand for Precision and Recall of the positive class (anomaly) while PN and RN stand for Precision and Recall of the negative class (not anomalous situation).

6   Numerical Results and Discussion

Generally, in the considered industrial setup, it would be more desirable to avoid undetected anomalies (which may indicate some error in the robot or some external issue in the predefined trajectory) than obtaining a false positive. Hence, all the figures in this section shows the algorithm performance in terms of the PF_1 metric.

6.1   Impact of Sampling and Noise Averaging

To evaluate the impact of both sampling and noise averaging, we consider an LIS compounded by $M = 128 \times 128$ antennas and a spacing $\Delta s = \lambda/2$ for the $\Delta d = 50$ cm anomalous route.

For our particular case, $N'_s \in \{1000, 500, 100\}$. Then $\forall \, \mathbf{p}_j$ we use $S = \frac{N'_s}{N_s}$ samples for obtaining N_s S-averaged samples for training the algorithm, being still $T = N_p \times N_s = 3670$. Note that the number of samples N'_s would depend on the sampling frequency and the second order characterization of the channel, i.e., the channel coherence time and its autocorrelation function.

Figure 4 shows the performance of the system when using non-averaged samples and averaged ones respectively. The blue line represents the system when non-averaged data is being used. When the noise contribution is non-negligible in the interval $\overline{\gamma} \in [10 \text{ dB}, 0 \text{ dB}]$, the detection performance presents a significant drop. Thanks to the averaging, results are significantly improved, even in the critical interval. As expected, when noise level is higher, more samples are needed to preserve the pattern by averaging, being $N'_s = 1000$ the one which yields a better performance. For the following discussions, this sampling strategy will be used, meaning we are using $S = 100$ extra samples.

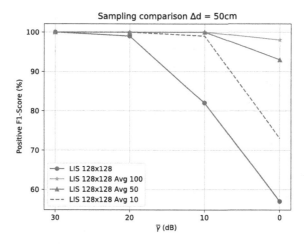

Fig. 4. PF_1 score averaged noise vs non-averaged.

6.2   Impact of Antenna Spacing

To evaluate the impact of inter-antenna distance, we fix the aperture to 5.44×5.44 m, we assess the performance in both $\Delta d = 50/10$ cm, and we analyze different spacings with respect to the wavelength ($\lambda/2$, λ and 2λ).

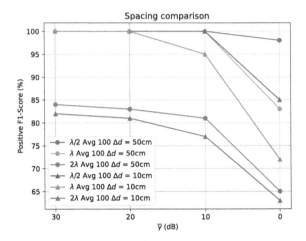

Fig. 5. PF_1 score antenna spacing

The performance results for the distinct configurations are depicted in Fig. 5. As observed, the spacing of 2λ—which is far from the concept of LIS—is presenting really inaccurate results showing that the spatial resolution is not enough. We can conclude that the quick variations along the surface provide important information to the classifier performance. Besides, this information becomes more important the lower the distance between the routes is. The performance drop due to the closer distances among the routes is related to the pattern classification. The closer the routes are, the more similar the pattern is making more challenging to perform the detection. However, reducing the antenna spacing even more can improve the performance when routes are even closer. What is more, the effect of antenna densification for a given aperture is highlighted and it can be seen that the lowest spacing leads to the best results.

6.3   LIS Aperture Comparisons

In this case LIS with different apertures have been evaluated. The spacing is fixed to $\lambda/2$.

Looking at Fig. 6, the aperture plays a vital role in the sensing performance. Increasing the number of antennas leads to a higher resolution image, being able to capture the large-scale events occurring in the environment more accurately. Note the usage of incoherent detectors is yielding to a good performance when the aperture is large enough. The key feature for this phenomena is the LIS pattern spatial consistency, i.e., the ability of representing the environment as a continuous measurement image.

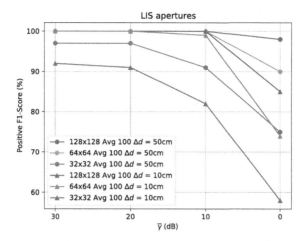

Fig. 6. Different LISs apertures comparison

7   Conclusions

We have shown the potential of LIS for sensing the environment, being able to provide high resolution radio images of the propagation environment that can be processed by existing and versatile solutions in the context of computer vision algorithms. This sensing technique, which we consider appropriate to refer to as holographic sensing, arises as a robust solution to capture the large scale events of a target scenario, with the inherent advantage that the received signal phase does not need to be estimated. The combined usage of both LIS and machine learning algorithms may be potentially used in the context of cognitive radio and multiuser massive MIMO as a support technology to enhance the performance of these systems.

References

1. Andrews, J.G., et al.: What will 5G be? IEEE J. Sel. Areas Commun. **32**(6), 1065–1082 (2014)
2. Hu, S., Rusek, F., Edfors, O.: Beyond massive MIMO: the potential of data transmission with large intelligent surfaces. IEEE Trans. Signal Process **66**(10), 2746–2758 (2018)
3. Basar, E.: Transmission through large intelligent surfaces: a new frontier in wireless communications. In: EuCNC: European Conference on Networks and Communications, pp. 112–117. IEEE (2019)
4. Dardari, D.: Communicating with large intelligent surfaces: fundamental limits and models. IEEE J. Sel. Areas Commun. **38**(11), 2526–2537 (2019)
5. Björnson, E., Sanguinetti, L.: Power scaling laws and near-field behaviors of massive MIMO and intelligent reflecting surfaces. IEEE Open Commun. Soc. **1**, 1306–1324 (2020)

6. Wang, H., Zhang, D., Wang, Y., Ma, J., Wang, Y., Li, S.: RT-Fall: a real-time and contactless fall detection system with commodity WiFi devices. IEEE Trans. Mobile Comput. **16**(2), 511–526 (2016)

7. Pu, Q., Gupta, S., Gollakota, S., Patel, S.: Whole-home gesture recognition using wireless signals. In: Proceedings of the 19th Annual International Conference on Mobile Computing & Networking, pp. 27–38 (2013)

8. Zhao, Y., Patwari, N., Phillips, J.M., Venkatasubramanian, S.: Radio tomographic imaging and tracking of stationary and moving people via kernel distance. In: 2013 ACM/IEEE International Conference on Information Processing Sensor Networks (IPSN), pp. 229–240. IEEE (2013)

9. Adib, F., Kabelac, Z., Mao, H., Katabi, D., Miller, R.C.: Real-time breath monitoring using wireless signals. In: Proceedings of the 20th Annual International Conference on Mobile Computing Networking, pp. 261–262 (2014)

10. Zhao, M., et al.: Through-wall human pose estimation using radio signals. In: Proceedings of the IEEE Conference on Computer Vision and Pattern Recognition, pp. 7356–7365 (2018)

11. Wilson, J., Patwari, N.: Radio tomographic imaging with wireless networks. IEEE Trans. Mobile Comput. **9**(5), 621–632 (2010)

12. Joung, J.: Machine learning-based antenna selection in wireless communications. IEEE Commun. Lett. **20**(11), 2241–2244 (2016)

13. Demir, O.T., Bjornson, E.: Channel estimation in massive MIMO under hardware non-linearities: Bayesian methods versus deep learning. IEEE Open J. Commun. Soc. **1**, 109–124 (2020)

14. Ma, X., Gao, Z.: Data-driven deep learning to design pilot and channel estimator for massive MIMO. IEEE Trans. Veh. Technol. **69**(5), 5677–5682 (2020)

15. Huang, H., Yang, J., Huang, H., Song, Y., Gui, G.: Deep learning for super-resolution channel estimation and DOA estimation based massive MIMO system. IEEE Trans. Veh. Technol. **67**(9), 8549–8560 (2018)

16. Syms, R.R.A.: Practical volume holography clarendon. Oxford **19902**, 125 (1990)

17. Pan, S.J., Yang, Q.: A survey on transfer learning. IEEE Trans. Knowl. Data Eng. **22**(10), 1345–1359 (2009)

18. Sarkar, D., Bali, R., Ghosh, T.: Hands-On Transfer Learning with Python: Implement Advanced Deep Learning and Neural Network Models Using TensorFlow and Keras. Packt Publishing Ltd., Birmingham (2018)

19. Chollet, F., et al.: Keras (2015). https://keras.io

20. Bishop, C.M.: Pattern Recognition and Machine Learning. Springer, New York (2006)

21. Simonyan, K., Zisserman, A.: Very deep convolutional networks for large-scale image recognition. arXiv preprint arXiv:1409.1556 (2014)

22. Winprop - Altair Engineering Inc. https://www.altairhyperworks.com/winprop

23. Anguita, D., Ghio, A., Ridella, S., Sterpi, D.: K-fold cross validation for error rate estimate in support vector machines. In: DMIN, pp. 291–297 (2009)

24. Roy, P., Ghosh, S., Bhattacharya, S., Pal, U.: Effects of degradations on deep neural network architectures. arXiv preprint arXiv:1807.10108 (2018)

25. Moeslund, T.B.: Introduction to Video and Image Processing: Building Real Systems and Applications. Springer, Heidelberg (2012). https://doi.org/10.1007/978-1-4471-2503-7

26. Powers, D.M.: Recall & precision versus the bookmaker. In: International Conference on Cognitive Science (2003)

Business Models and Spectrum Management

Scalability and Replicability of Spectrum for Private 5G Network Business: Insights into Radio Authorization Policies

Pekka Ojanen[1]([☒]) and Seppo Yrjölä[2,3]

[1] Co-Worker Technology Finland Oy, Turku, Finland
pekka.ojanen@co-workertech.com
[2] Nokia, Oulu, Finland
[3] Centre for Wireless Communications, University of Oulu, Oulu, Finland

Abstract. New spectrum bands are being released to respond to the growing need for locally deployed industrial and private networks. This calls for new licensing schemes and spectrum sharing approaches. New challenges are faced from the ever-increasing variety of released spectrum bands with different technical and operational requirements and the increasing fragmentation of spectrum management approaches. While the standardization is progressing and technical solutions are developed for the new networks, less attention has been paid to the radio product related regulation, including equipment authorization frameworks. With roots in engineering, policy and economics, this paper looks through the lenses of business model framework at scenarios of wireless equipment authorization related to novel spectrum management approaches. This paper provides an overview of radio authorization for mobile communication networks and develops a conceptual framework to depict and analyze authorization policies. The results indicate the strong impact the authorization frameworks have on the scalability and replicability of the business. This calls for novel models of governance and regulation that highlight utilization of harmonized, widely employed frequency bands and the associated technical requirements.

Keywords: Business model · Cognitive radio · Private networks · Radio authorization regulation · Spectrum policy · Spectrum sharing · 5G

1 Introduction

The deployment of 5th generation (5G) mobile networks aims at significant increase of the efficiency and productivity of industries, logistics, agriculture and other vertical sectors. Key technical elements will be the wide employment of Internet of Things (IoT), possibility for very low network latencies and highly increased transmission capacity [1]. Depending on the use cases and their requirements there may be a need to allow the enterprises and industries to deploy private networks without relying on communication service providers networks [2]. The industrial use cases can be significantly different

© ICST Institute for Computer Sciences, Social Informatics and Telecommunications Engineering 2021
Published by Springer Nature Switzerland AG 2021. All Rights Reserved
G. Caso et al. (Eds.): CrownCom 2020, LNICST 374, pp. 141–157, 2021.
https://doi.org/10.1007/978-3-030-73423-7_11

from those supported by public mobile networks, especially related to requirements on guaranteed network performance, trustworthiness, security and privacy [3]. If an enterprise intends to deploy a private network as a local operator [4], it needs an authorized timely access to affordable quality spectrum [5, 6].

To date only a limited number of countries have released dedicated spectrum for private networks. Required spectrum availability, sharing and authorization models for local 5G networks were discussed in [7], locally and temporarily shared spectrum in [8] and micro licensing and the associated regulatory framework in [9]. Reference [10] reviewed several frameworks based on local deployments utilizing geographical sharing and cognitive radio techniques like the US three-tiered Citizens Broadband Radio Service (CBRS) and the license exempt 5 GHz shared Radio Local Area Network (RLAN) bands. Furthermore, it introduced a method for assessing spectrum management approaches applicable to private industrial networks and assessed the suitability of the recent frameworks below 5 GHz to use cases of private industrial networks.

In addition to valid use cases and access to affordable spectrum, the timely availability of radio equipment is essential. Applicable radio equipment needs to be authorized in order be placed on the market and be used. Information on radio equipment authorization is mainly discussed and available as standards and as regulatory documents, e.g., [11]. Engineering overview of type approval is discussed, e.g., in [12–14] and national authorization frameworks are presented by commercial entities in [15–17]. To the best of the authors' knowledge there is no preceding research assessing the radio authorization approaches in the context of locally deployed private networks and novel spectrum regulation frameworks.

This cross-disciplinary paper looks at business aspects related to novel spectrum management approaches, particularly scenarios of wireless equipment authorization through the lenses of business model framework [18] focusing on scalability [19], replicability [20], and sustainability [21]. Scalability can be categorized based on the five mutually exclusive causal factors: technology, cost and revenue, legal regime adaptability, network effects and user orientation [22]. Replicability, on the other hand focuses on how firms replicate their successful business models across geographical markets [20, 23]. The research discussing localization of the mobile communication services has recently emerged in the business model literature. Matinmikko et al. [4] and Ahokangas et al. [24] defined the local operator concept, stakeholders, and related business models. Regulatory requirements were explored in [25], and the challenge in spectrum management revealing a new level of complexity that stems from the variety of spectrum bands and spectrum access models with different levels of spectrum sharing and cognitive radio techniques in [26]. Business opportunities for local private network operators found in [4] were hosting for mobile network operators (MNOs), secure and private local service provisioning for industrial enterprises, offering differentiating service level agreements, and expanding service to data and application governance [4].

This study widens and enhances the earlier research via focusing on radio product availability and examines radio equipment authorization approaches as antecedents to private industrial 5G networks deployment. In order to assess the scalability and replicability of the radio authorization approaches, the radio product test and certification processes and regulation in the 27 European Union (EU) countries and in 67 other countries have been studied. The countries were chosen based on their relevance for private mobile network business, cognitive radio spectrum management approaches and availability of detailed regulatory data. Special attention was paid to the US and the EU/European Telecommunications Standards Institute (ETSI) radio authorization frameworks applicable in multiple countries globally. This study is based on the direct communication with selected national regulatory authorities, and data collected from governmental web pages.

The rest of the paper is organized as follows. In Sect. 2, the theoretical foundation and key concepts are presented. The radio authorization frameworks are presented and discussed in Sect. 3, and conclusions are drawn in Sect. 4.

2   Theoretical Foundation and Key Concepts

This section discusses the key concepts and frameworks utilized in the study: concept of business model, radio spectrum requirements for private 5G networks and the radio equipment authorization processes as depicted in Fig. 1.

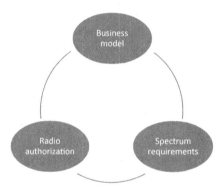

Fig. 1. Key concepts and frameworks.

2.1   Business Model Framework

The contemporary business model research has evolved towards value creation, value sharing and value capture discussion [27] from traditional activity view [28]. Furthermore, recent studies stemming from ecosystem perspective [18] do not presume a focal firm and sees business model to "*create value through the exploitation of business opportunities*". In addition to value and opportunity, the exploration and exploitation of competitive advantages form strategic business choices of company [27, 29–33]. In studies analyzing the antecedents of the prosperous business, three strategic implication were found: capabilities to *scale* [19], *replicate* [20], and *sustain* [21]. This paper focused on analyzing radio product regulation and its relation to scalability and replicability as growth related factors of business models. Scalability can be defined as the ability of a network to be extended in a capable flexible manner [34] where the augmentation of resources results in increasing returns [35]. Organizational flexibility [36] allows agility to cope with changes in the industrial environment such as competitive and regulative landscape or macro-economic pressures [36]. Scalability can be categorized based on the five mutually exclusive causal factors: technology, cost and revenue, legal regime adaptability, network effects and user orientation [22]. Replicability, on the other hand focuses on how firms replicate their successful business models across geographical markets [20, 23] in order to maximize value [37]. The replication strategy stems from company's dynamic capabilities to choose, refine, transfer and maintain the necessary elements of their business model components to allow successful replication in applicable geographies [23].

2.2 Private 5G Network Spectrum Requirements

For the fourth generation (4G) public mobile networks, radio frequencies were generally allocated and assigned utilizing competitive awarding, i.e. via auctions, and the same approach is most widely used with the 5G spectrum release [38, 39]. However, several countries have already introduced regulatory frameworks aimed at deployment of individually authorized, locally deployed private 4G or 5G networks [40]. In most cases dedicated spectrum has been released under local licensing, allowing for local network deployments through basic geographical sharing, assuming that in each location there is only one network. However, there are also regulatory frameworks utilizing specific technical and operational cognitive radio requirements to facilitate sharing with incumbents and other users, for example in the US for the CBRS band [41], in the UHF broadcasting band for TV White Space (TVWS) [42] use and for the European Licensed Shared Access (LSA) concept [43]. As the demand for spectrum for local deployments increases, spectrum management frameworks with operational requirements related to more dynamic spectrum access and cognitive technical capabilities are to be defined and employed.

Many industrial use cases have very specific technical and operational requirements which call for specific network implementations that are best realized though deployment of a private network [44]. The spectrum related preferences of private network deployments were studied in [10] and [45]. The preferences aim at gaining from economies of scale for lowering the equipment cost and at minimizing the spectrum and regulatory costs. In case of public mobile networks, the large market size and the utilization of harmonized 3GPP spectrum bands allows for economies of scale for equipment, while for private networks the case is very different due to fragmented bands, specific technical and operational requirements, as well as due to limited in-country market size. Furthermore, the fragmentation is reflected in fragmentation of the radio equipment authorization requirements. Therefore, approaches increasing the geographical availability (replicability) and applicability (scalability) of the authorization frameworks have become important antecedents of successful business. The proposed replicability and scalability assessment framework comprised of the following elements and preferences, adapted from [45], as shown in Tables 1 and 2.

Table 1. Replicability related preferences.

Replicability (Geographical availability)	
Characteristic	Preference
Band availability	Widely available for ecosystem and economies of scale
Radio authorization	Widely utilized framework employed (FCC, EU/ETSI)

Table 2. Scalability related preferences.

Scalability (Applicability)	
Characteristic	Preference
Bandwidth	Sufficient for wideband applications
Network deployment	Both indoor and outdoor
Max transmit power	Sufficient for indoor and/or outdoor deployments
Interference protection	Yes
Co-existence mechanisms	No complex technical or restrictive operational requirements
Regulatory certainty	Predictable framework and schedule, fast time-to-air
Spectrum cost/pricing	Fee, administrative one-off or annual fee, per license/area
Standards maturity	Standardizations process mature in IEEE and/or 3GPP
Radio availability	Yes

Especially the interference protection and co-existence mechanism related technical and operational requirements are closely related to possible spectrum sharing, employed cognitive elements and dynamic spectrum use. Currently, radio equipment for the US CBRS band, the LSA band, TVWS band or any other band where cognitive capabilities or band specific requirements are required must conform to specific technical requirements in the compliance testing.

2.3 Radio Equipment Authorization

In this section radio authorization processes used in the analysis are discussed in more details. Only locally authorized radio products are allowed to be placed on the market and be used [11, 46]. The radio equipment authorization is achieved by verified conformance to local requirements and regulations resulting in a certification issued by the local regulator or by a body authorized by the regulator. The radio authorization requirements and the approval and certification procedures vary by region or country. This must be taken into account as there may be a need to go through extensive and very specific type approval testing process in the target country to obtain certification before the equipment can be placed on the market and taken into use. Moreover, the equipment authorization process can be very complex and time consuming, especially in case novel spectrum management frameworks are utilized.

Product authorization may be achieved through Type Approval (TA) or by Supplier's Declaration of Conformity (SDoC). Most countries require a TA of radio products to demonstrate the conformance. TAs may comprise of laboratory testing against the national technical requirements/standards and certification including product labeling and may require in-country laboratory tests or be partly or completely based on submitted documents. It is also possible that a document-based application process must be complemented by submission of product samples and verification of their performance. In the EU and certain individual countries, SDoC allows placing the products on the market.

Mutual Recognition Agreements (MRAs) are agreements established between countries for the purpose of mutual recognition of conformity assessment of regulated products. The benefits of the MRAs arise from removal of duplicated inspection or certification. Where a product intended for two markets may still have to be assessed twice (when technical requirements or standards are different), the assessment cost will be lower when carried out by one body. Furthermore, the time to market and business is reduced since contacts between the manufacturer and the single conformity assessment body, and a single assessment, speed up the process. Even where the underlying regulations are harmonized, for example because they refer to the same international standard, the need for recognition of certificates remains, and in such cases the benefit will be clear: the product is assessed only once against the commonly accepted standard instead of twice. Wireless product authorizations require typically demonstrating compliance also with requirements in other areas than radio/wireless, such as compliance with Electromagnetic Compatibility (EMC), energy efficiency, safety, and environmental requirements. Based on the global radio authorization method data, the following framework was developed to characterize radio product authorization, as summarized in Table 3.

Table 3. Radio product authorization characterization.

Product authorization characteristic	
Scheme	Employed approval scheme (Type Approval, Supplier's Declaration of Conformity…) and description of required process, including possible requirement for laboratory tests and required documentation
Local presence	Requirement for local applicant/representative
Process costs	Costs and payment instructions
Process timing	Prior information on duration of the authorization process
Technical requirements	Applicable standards
Test facilities	Requirement for in-country testing or usage of accredited in-country or international test laboratories
Foreign test applicability	Mutual recognition agreements (applicability of foreign test results, especially FCC and EU related tests)
Marking and documentation	Requirements for product labeling and documentation
Authorization duration	Validity period of authorization, renewal conditions

3   Analysis of Radio Authorization Frameworks

This section presents collected data and analyses the main aspects of radio authorization frameworks in the European Union covering Austria, Belgium, Bulgaria, Croatia, Cyprus, Czechia, Denmark, Estonia, Finland, France, Germany, Greece, Hungary, Ireland, Italy, Latvia, Lithuania, Luxembourg, Malta, Netherlands, Poland, Portugal, Romania, Slovakia, Slovenia, Spain, Sweden, in the US and in 66 other countries, covering Andorra, Angola, Argentina, Australia, Bahrain, Bosnia and Herzegovina, Botswana, Brazil, Cambodia, Canada, Chile, China, Colombia, Egypt, Equatorial Guinea, Gabon, Guatemala, Hong Kong, Iceland, India, Indonesia, Israel, Ivory Coast, Japan, Jordan, Kenya, South Korea, Kuwait, Lebanon, Liechtenstein, Macau, Malaysia, Mexico, Moldova, Montenegro, Morocco, Mozambique, Namibia, New Zealand, Nigeria, North Macedonia, Norway, Panama, Peru, Philippines, Puerto Rico, Qatar, Russia, Rwanda, Saudi Arabia, Serbia, Singapore, South Africa, Taiwan, Tanzania, Thailand, Tunisia, Turkey, Uganda, Ukraine, United Arab Emirates, United Kingdom, Venezuela, Vietnam and Zimbabwe.

3.1   European Radio Authorization Framework

In the EU, the Radio Equipment Directive (RED) provides the essential requirements for placing radio equipment on the market [47]. The essential requirements address safety and health, electromagnetic compatibility and the efficient use of the radio spectrum. All radio products in scope of this directive and placed on the EU market must be compliant with the RED. Only compliant radio equipment may be sold and used in the EU. The essential performance requirements and use of radio spectrum are laid down in Articles 3.2 and 3.3 of the RED [47]. The RED allows manufacturers to self-declare that their equipment meets the applicable ETSI harmonized standards, which at the same time indicates compliance with the essential requirements of the RED, and to affix the CE mark, so the equipment can be placed on the market in those countries where the harmonized standards apply. If there are applicable harmonized standards, there is no requirement for TA testing by the regulator or associated accredited bodies. But if there is no applicable harmonized standard then manufacturers can use any technical specification to demonstrate that the radio equipment complies with all necessary requirements, upon their own responsibility. The results must, however, be certified by a Notified Body, if alternative specifications than harmonized standards are applied. The EU has MRAs currently in force with Australia, New Zealand, the United States, Canada, Japan and Switzerland.

In Europe, seven countries, Finland, France, Germany, the Netherlands, Norway, Sweden and the UK, have designated spectrum for private 4G or 5G networks, and several countries are considering doing so. Even if a variety of band is employed, the common radio authorization framework applies. However, there are country specific technical requirements, for example in the UK, for the utilization of the 2390–2400 MHz, 3800–4200 MHz and 24250–26500 MHz bands [48], which limits the replicability of the equipment authorization related to those bands.

3.2 US Radio Equipment Authorization Framework

All electrical equipment marketed in the USA requires authorization. Intentional radiators, including radio transmitters, are required to be tested at an authorized test laboratory and the technical file reviewed by an independent body before a certificate is granted and details listed on the Federal Communications Commission (FCC) website. The FCC has two different approval procedures for equipment authorization: Certification and Supplier's Declaration of Conformity. The required procedure depends on the type of equipment being authorized as specified in the applicable rule part. In some instances, a device may have different functions resulting in the device being subject to more than one type of approval procedure. Telecom products having a radio transmitter, such as mobile phones or RLAN equipment are to be certified. Technical rules are generally specified in applicable parts of [49], and the administrative rules for equipment authorization, including the process are described in [50]. The US has MRAs for conformity assessment of telecommunications equipment in place with Asian Pacific Economic Cooperation (APEC), Inter-American Telecommunications Committee of the Organization of American States (CITEL), the EU, Israel, Japan and Mexico.

The band 3550–3700 MHz has been made available in the US for CBRS [41] and it is expected to become the main band for enterprise networks. The band accommodates currently the prioritized incumbent users (tier 1) protected from the potential interference originated from the lower tiers. The rules provide two paths for local industrial spectrum: tier 2 Priority Access License (PAL) and tier 3 General Authorized Access (GAA). The FCC allows the PAL licensees to lease their spectrum, which is within their PAL area, but beyond their deployment coverage. For example, in an industrial area a PAL holder may lease one or more of their channels to industrial enterprises. The General Authorized Access (GAA) users can access the portions of the CBRS band in 3550–3650 MHz that are unused by the incumbents and the PAL users and the portions of the band in 3650–3700 MHz that are unused by the incumbents. The GAA usage is also suitable for enterprise usage in case the spectrum availability is not critical. There are CBRS band specific requirements for radio equipment. Specific requirements also extend to authorization of other related network elements, as applies for example to the Environmental Sensing Capability (ESC) entities and Spectrum Access System (SAS) in the CBRS band [41]. Employment of specific requirements limits the scalability of the authorization frameworks and creates new challenges to authorization of radio products for private networks.

3.3 Comparison of FCC and EU Radio Product Authorization Processes

Table 4 presents and compares the FCC and EU radio product authorization processes. The major difference is that the FCC process builds on testing at authorized test laboratory and certification granted based on application, while the EU process builds on manufacturer's self-testing and self-declaration of conformity.

3.4 Radio Authorization Frameworks in Selected Countries

In addition to the EU and US radio authorization frameworks addressed in the previous section, approaches in 66 selected countries were analyzed. Radio authorization

Table 4. EU and FCC product authorization process comparison.

Radio product authorization	FCC	EU
Approach	Testing and certification	Self-testing and Supplier's Declaration of Conformity
Preparations	Determine applicable rules, e.g. in CFR Title 47	Determine applicable ETSI Harmonized Standards (ETSI HS)
Testing	Compliance testing at an authorized testing lab	In case ETSI HS covers product: manufacturer's own tests are sufficient. In case ETSI HS does not (fully) cover product: tests required at Notified Body
Approval	Apply for Certification	Prepare Supplier's Declaration of Conformity
Labeling	Attach FCC Certification Number, include compliance info in manual	Label product (CE mark), include compliance info in product documentation
Outcome	Product ready for US market	Product ready for EU market

frameworks in nine countries together with examples of relevant spectrum bands are presented in this section, as an example. Countries were selected based on private network opportunities, novel spectrum management approaches and different authorization frameworks.

Australia

In Australia spectrum has been reserved for private network deployments for example in the 5.8 GHz band and in 24 GHz. The 5.8 GHz band is available under general authorization for private networks. The requirements are partly similar to the FCC requirements, but partly country specific. A portion of the 24 GHz band will be available as licensed band, and another portion under general authorization, without a license, both suitable for local network deployments. The technical requirements for the 24 GHz band would be at least partly country specific.

The radio equipment authorization in Australia is based on Declaration of Conformance. Compliance with Australian ACMA technical standards for telecommunications customer equipment, radiocommunications devices, electromagnetic radiation, and electromagnetic compatibility is required. The process is document based and no in-country test is required. A local representative is required, either the manufacturer, importer or agent. Evidence of transmitter compliance to ACMA technical standards may be demonstrated by providing a complete ETSI or FCC test report. The Australian importer or supplier is required to make the declaration of conformity and hold a product description and compliance records. Compliant products should also be labelled with the Regulatory Compliance Mark (RCM). The validity period of the authorization is unlimited. The wireless and EMC requirements are similar in Australia and New Zealand, therefore the approval applied in Australia also indicates similar compliance in New Zealand and vice versa [51].

China

Spectrum management in China follows largely the traditional way to make the mobile bands available for major MNOs and countrywide deployment of public mobile networks. The recent development in Chinese Belt and Road Initiative (BRI) is exporting China originated frequency and radio standard variants into several South American and African counties targeting in particular the utility verticals, e.g., on 200 MHz and 450 MHz spectrum bands.

All radio transmitter devices in China require Radio Type Approval (RTA) issued by the State Radio Monitoring Centre of the Ministry of Information (SRRC) [52]. The approval process includes in-country testing at the State Radio Monitoring Centre (SRMC) and a certification application to the SRRC. A local applicant/representative is required. Approved radio equipment must be marked with a certificate number (CMIIT ID) that must appear on products or labels. The validity period of the authorization is 5 years, and one renewal is possible. A Network Access License (NAL) is required for all telecommunications equipment that connects to public network. Safety and EMC testing and The Restriction of Hazardous Substances (RoHS) certification is also required.

Hong Kong

The band 27.95–28.35 GHz has been made available on a geographically shared basis for locally provided, innovative wireless broadband services targeted towards specific groups of users. The deployed networks are based on the 5G or other advanced mobile technologies. This is a country specific spectrum opportunity for deployment of private networks.

Radio equipment shall meet the minimum requirements of the Hong Kong Communications Authority (HKCA) [53]. The requirements are described normally in the form of HKCA specification. Radio equipment which has been evaluated against the relevant HKCA specification and is compliant with the requirements may be authorized. Local certification bodies and foreign certification bodies have been accredited or recognized to provide type-approval of radio equipment. The certification is done under the Hong Kong Telecommunications Equipment Evaluation and Certification (HKTEC) Scheme. Under the scheme, radio equipment is classified under the Voluntary Certification Scheme (VCS) or Compulsory Certification Scheme (CCS). Radio equipment classified under VCS can be used or marketed in HK without type approval. Generally, mobile equipment is under VCS, but for example LTE base stations are under CCS.

India

Spectrum allocation in India follows largely the traditional approach where mobile bands are made available for MNOs. However, the 5.8 GHz band allows for deployment of private networks, and the technical requirements are largely similar to the FCC requirements. The same 5.8 GHz band is available in more than ten countries globally.

All wireless equipment shall be type approved [54]. The Wireless Planning and Coordination Wing, Ministry of Communications and IT, Department of Telecommunications (WPC) issues approval of radio devices operating in licensed and unlicensed frequency bands. For radio equipment operating in licensed frequency bands the importers, dealers and users of the equipment must obtain licenses from WPC. An application with essential documents and test reports must be sent to the relevant agencies by the manufacturer. India-based representative is required. Generally, the approval is document based, i.e., based on application and review of foreign standard test reports and approval certificates. FCC or EU/ETSI test documentation may be utilized. Online filing is possible for all types of license applications, followed by hard copy submission of the application. Certain products must be tested in-country. Approved equipment must be labeled with TEC label. The validity of the authorization is unlimited.

Japan

Japan has made the band 28.2–28.3 GHz available for locally deployed 5G networks. The eligibility is connected to the land ownership or right of use. The existing MNO's are not granted access to the band. Both indoor and outdoor use are allowed. The aim is to make adjacent band, i.e. 28.3–29.1 GHz available in late 2020, when the sharing studies on the conditions allowing sharing with incumbents are completed.

Testing radio equipment to be used in Japan to conform to Japan's radio standards may be conducted by properly recognized laboratories. and they must be certified by one of several private MIC Recognized Certification Bodies (RCB) [55]. The EU – Japan MRA enables appointed certification bodies in the EU to grant approval for certain radio equipment enabling fast access to the Japanese market. The details of the approval process depend on the category in which the product falls under. No in-country testing is required and no local applicant is required. Approved equipment must be labeled with MIC certificate. The validity of the authorization is unlimited.

South Africa

South Africa follows largely the traditional operator model, where most of the mobile bands are assigned to public mobile use. The 5.8 GHz band is available for Fixed Wireless Access (FWA) use and satellite spectrum in 2.4 GHz is being made available for private terrestrial use, similarly to several African countries. Similarity between the South African and European spectrum allocations and radio equipment authorization documentation will be a benefit if the regulator ICASA designates spectrum for private use.

All radio equipment is subject to type approval by ICASA [56]. In many cases the ICASA will issue approval based on review of foreign standard test reports that demonstrate compliance with South African technical standards. The RED directive applies, and EU/ETSI test documentation may be utilized as ICASA accepts reports on product tests conducted based on the relevant EN standards, if the testing has been performed at an accredited test lab. The applicable technical standards are found in the technical regulations as defined in the TA regulations. Applications can be submitted by manufacturers, importers, distributors, and any South African registered company. The TA certificate will only be issued to South African registered companies. The approval usually takes 30 days, from submission of all necessary documents. Equipment must be labeled with ICASA type approval label. The validity of the approval is unlimited.

Canada

The Wireless Broadband Service is allowed to use the frequencies 3650–3700 MHz in Canada. The deployed technologies include WiMax and LTE, adapted to comply with the band specific requirements. Both fixed and mobile network deployments are allowed. In practice, the band is largely used for local industrial networks.

The regulator ISED requires certification for telecommunications terminal equipment and radio devices [57]. The product must comply with Canadian regulations and ISED standards. Applications for certification are submitted to recognized certification body or ISED. A Canadian representative is required for certifications or registrations. Testing to demonstrate compliance with Canada's technical standards can be conducted at many recognized laboratories: globally there are 37 Test laboratories and 40 certification bodies recognized by ISED. Approved equipment must be labeled with product ID info and ISED registration number. The validity of the approval is unlimited.

Mexico

Mexico does not have licensed bands for enterprise use. However, the 5.8 GHz band is available under a general authorization. The allocation is similar to the US, but the technical requirements are partly country specific. Leasing spectrum from MNOs could be another option.

All radio products must be certified by the regulator IFT [58]. Compliance with the official Mexican standards and technical provisions is required. In-country testing in accredited laboratories is required, as well as a local representative. Submission of two samples of certain parts is required, as well as employed test software, product specifications and installation and operation manuals. The FCC approval only helps as provider of additional info. Testing and certification of each product and model is required separately. All approved products should be labeled with IFT label. Validity of provisional certification is one year, renewal is unlimited.

Argentina

The 450 MHz band has been made available for regional use out of the major cities to cover 373 areas in 15 provinces. This could be an opportunity for private networks in rural areas.

The regulator ENACOM requires that all radio products comply with the national requirements [59]. There are two approval processes Homologación and Codificación. The former is related to approval related to Argentine standards and the latter related to approval in case no local standard exists. Equipment testing must be performed in-country at an accredited test laboratory. Foreign test reports are not accepted, but they may be helpful as provider of additional info. An Argentinian representative (applicant) is required. The importer or distributor must be registered with ENACOM in order to obtain certification. The approved products must be labeled with a CNC ID. The validity period of the approval is 3 years.

3.5 Discussion

This section summarizes the findings of radio authorization frameworks addressed in this study and assesses their impact on scalability and replicability of the business.

Summary of Radio Authorization Frameworks Globally

In all countries only locally authorized radio products are allowed to be placed on the market and be used. The application process is typically country specific, and involves submission of forms and other national documents, often in the local language, interaction with different authorities, usually at least with the local regulator, and payments. Most countries require TA of radio products to demonstrate the conformance to national requirements. TAs may comprise of laboratory testing against the national technical requirements/standards and certification including product labeling. In-country laboratory tests may be required. The approval may be partly or completely based on submitted documents. The study shows that TA/certification is required in 40 countries. In the EU and certain individual countries, such as Australia, New Zealand, Hong Kong, and Lebanon, Supplier's Declaration of Conformity allows placing the products on the market. Applicable standards are usually specific national standards or specifications, which may refer to international standards. A local applicant is required at least in 30 non-EU countries, in-country laboratory tests are required in 14 non-EU countries. Product labeling is required in 38 countries and the EU27. The validity period of certification/approval is unlimited in the EU and 36 other countries, while in 27 countries the validity periods very between 1 and 5 years.

Replicability and Scalability of Radio Authorization Frameworks

In many cases the national standards refer to a few key technical and operational requirements, which usually have similarities to ITU-R requirements. For example, for ITU Region 1 countries both the spectrum allocations [60] and the spectrum related requirements are very similar, thus the technical requirements may be similar to European requirements and the results of product tests based on ETSI standards are valid in many countries also outside of Europe, especially in Africa and in the Middle East, all belonging to ITU-R Region 1. The same applies to the ITU Region 2, especially in the Latin-American countries the national spectrum use, and the related requirements have similarities with the US spectrum use and FCC requirements. Therefore, the national technical requirements in many countries globally have similarities to or are identical with ETSI or FCC frameworks and requirements. On the other hand, in most countries the ETSI or FCC product authorizations are not sufficient as such for local product authorization. However, in the authorization process it is possible to utilize EU/ETSI documentation in 38 countries outside of EU and the FCC documentation in 8 countries. A pre-requisite for this is some degree of commonality between the local technical requirements and applicable standards and ETSI/FCC requirements. Figure 2 shows the global replicability of FCC and EU authorization test results in the countries addressed in this study.

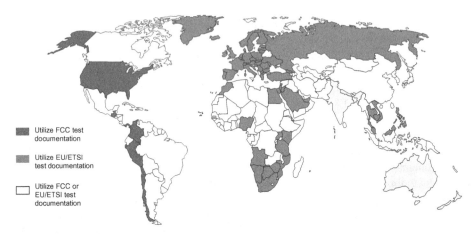

Fig. 2. Global replicability of FCC and EU/ETSI radio product authorization test results.

4   Conclusions

To date, only a few countries have introduced novel spectrum management frameworks to allow local private LTE or 5G networks to access individually authorized exclusive spectrum. Furthermore, the spectrum bands, allowed radio standards and the technical and operational requirements vary country by country. In addition to timely access to affordable spectrum, the key element for implementation of any private mobile network is availability of radio equipment that meet the use case requirements and business needs. The mandatory local authorization of radio products is achieved by verified conformance to local requirements and regulations, certified by the local regulator or by a body designated by the regulator or through supplier's self-declaration. Moreover, as the use cases and the markets of private networks are fragmented, and limited in size, the diversity of the product authorization frameworks is an issue, and the replicability and scalability of existing frameworks become important.

Extant research on radio authorization in the private network context is still scarce, and this paper represents the first attempt to depict the impacts on business. Utilization of harmonized, widely employed frequency bands and the associated technical requirements widen the replicability of the radio authorization frameworks. This can be enhanced by mutual recognition agreements between the countries. By employing similar technical and operational requirements to those of the FCC or EU, the scalability and replicability of the radio authorization frameworks can be further increased, leading to economies of scale.

Variety of spectrum bands and fragmentation will keep increasing in 6G, which calls for novel approaches in spectrum management and product authorization. With rapid technology developments and business driven needs the timely network deployment is essential and calls for more dynamic and forward-looking radio product authorization strategies. Since the 5G evolution towards 6G may be expected to be decisive for the competitiveness and sustainability of our economies, this paper sees the key future research challenge to be as follows: what kind of novel regulation, standardization, and governance models might emerge and be competitive in future 6G environments enabling successful scalable, replicable and sustainable business models both for nationwide mobile network operators as well as multiple local operators serving the long tail of enterprise, vertical and community customer needs.

Acknowledgment. The research has been supported by the Business Finland 5G Vertical Inte-grated In-dustry for Massive Automation (5G-VIIMA) program. The authors would like to acknowledge the support of the 6G Flagship programme at the University of Oulu.

References

1. Federal Ministry for Economic Affairs and Energy, Platform Industrie 4.0. https://www.pla ttform-i40.de/I40/Navigation/EN/Industrie40/WhatIsIndustrie40/what-is-industrie40.html. Accessed 18 Aug 2020
2. Zander, J.: Beyond the ultra-dense barrier: paradigm shifts on the road beyond 1000× wireless capacity. IEEE Wirel. Commun. **24**(3), 96–102 (2017)
3. 5G Alliance for Connected Industries and Automation (5G-ACIA): 5G for Connected Industries and Automation White Paper (2018)
4. Matinmikko, M., Latva-aho, M., Ahokangas, P., Yrjölä, S., Koivumäki, T.: Micro operators to boost local service delivery in 5G. Wirel. Pers. Commun. **95**(1), 69–82 (2017)
5. Ferrus, R., Sallent, O.: Extending the LTE/LTE-a business case: mission- and business-critical mobile broadband communications. IEEE Veh. Technol. Mag. **9**(3), 47–55 (2014)
6. Beltran, F.: Accelerating the introduction of spectrum sharing using market-based mecha-nisms. IEEE Commun. Stand. Mag. **1**(3), 66–72 (2017)
7. European Commission Radio Spectrum Policy Group, RSPG18-005 Strategic spectrum roadmap towards 5G for Europe: RSPG second opinion on 5G networks (2018)
8. Guirao, M.D.P., Wilzeck, A., Schmidt, A., Septinus, K., Thein, C.: Locally and temporary shared spectrum as opportunity for vertical sectors in 5G. IEEE Netw. **31**(6), 24–31 (2017)
9. Matinmikko, M., Yrjölä, S., Latva-aho, M.: Micro operators for ultra-dense network deploy-ment with network slicing and local spectrum micro licensing. In: The 2018 IEEE 87th Vehicular Technology Conference, Porto (2018)
10. Ojanen, P., Yrjölä, S.: Assessment of spectrum management approaches to private industrial networks. In: Kliks, A., Kryszkiewicz, P., Bader, F., Triantafyllopoulou, D., Caicedo, Carlos E., Sezgin, A., Dimitriou, N., Sybis, M. (eds.) CrownCom 2019. LNICST, vol. 291, pp. 277–290. Springer, Cham (2019). https://doi.org/10.1007/978-3-030-25748-4_21
11. FCC: Equipment authorization. https://www.fcc.gov/engineering-technology/laboratory-div ision/general/equipment-authorization. Accessed 18 Aug 2020
12. Rumney, M.: LTE and the Evolution to 4G Wireless - Design and Measurement Challenges, 2nd edn. Wiley, Hoboken (2013)
13. Mueck, M.D., Badic, B., Ahn, H., Bender, P., Choi, S., Ivanov, V.: Market access for radio equipment in Europe enabled by the radio equipment directive: status, next steps and implications. IEEE Commun. Mag. **57**(12), 20–24 (2019)
14. Knapp, J.P., Wall, A.: Streamlining the FCC equipment authorization process in response to changing global markets. In: Proceedings of Symposium on Electromagnetic Compatibility, pp. 1–10. Santa Clara, USA (1996)
15. iCertifi: Type Approval Certifications and Country Market Access Quick Links. https://www. icertifi.com/type-approval-certifications/. Accessed 18 Aug 2020

16. Compliance International: Countries. http://www.typeapproval.com/countries. Accessed 18 Aug 2020
17. Ingenieurbüro Lenhardt: Countries. https://ib-lenhardt.com/en/type-approval/. Accessed 18 Aug 2020
18. Amit, R., Zott, C.: Value creation in E-business. Strateg. Manag. J. **22**(6–7), 493–520 (2001)
19. Stampfl, G., Prügl, R., Osterloh, V.: An explorative model of business model scalability. Int. J. Prod. Dev. **18**(3–4), 226–248 (2013)
20. Aspara, J., Hietanen, J., Tikkanen, H.: Business model innovation vs replication: financial performance implications of strategic emphases. J. Strateg. Market. **18**(1), 39–56 (2010)
21. Schaltegger, S., Hansen, E., Lüdeke-Freund, F.: Business models for sustainability: origins, present research, and future avenues. Organ. Environ. **29**(1), 3–10 (2016)
22. Chrisman, J.J., Hofer, C.W., Boulton, W.R.: Toward a system for classifying business strategies. Acad. Manag. Rev. **13**(3), 413–428 (1988)
23. Winter, S.G., Szulanski, G.: Replication as strategy. Organ. Sci. **12**, 730–743 (2001)
24. Ahokangas, P., et al.: Generic business models for local 5G micro operators. IEEE Trans. Cogn. Commun. Netw. **5**(3), 730–740 (2019)
25. Matinmikko, M., Latva-aho, M., Ahokangas, P., Seppänen, V.: On regulations for 5G: micro licensing for locally operated networks. Telecommun. Policy **42**(8), 622–635 (2018)
26. Matinmikko-Blue, M., Yrjölä, S., Ahokangas, P.: Spectrum Management in the 6G Era: Role of Regulations and Spectrum Sharing. The 2nd 6G Wireless Summit, Levi, Finland (2020)
27. Zott, C., Amit, R., Massa, L.: The business model: recent developments and future research. J. Manag. **37**(4), 1019–1042 (2011)
28. Onetti, A., Zucchella, A., Jones, M., McDougal-Covin, P.: Internationalization, innovation and entrepreneurship: business models for new technology-based firms. J. Manage. Governance **16**(3), 337–368 (2012)
29. Morris, M., Schindehutte, M., Allen, J.: The entrepreneur's business model: toward a unified perspective. J. Bus. Res. **58**(6), 726–735 (2005)
30. Teece, D.: Business models, business strategy and innovation. Long Range Plan. **43**(2–3), 172–194 (2010)
31. McGrath, R.: Business models: a discovery driven approach. Long Range Plan. **43**(2–3), 247–261 (2010)
32. Gomes, J.F., Iivari, M., Pikkarainen, M., Ahokangas, P.: Business models as enablers of ecosystemic interaction: a dynamic capability perspective. Int. J. Soc. Ecol. Sustain. Dev. **9**(3), 1–13 (2018)
33. Osterwalder, A., Pigneur, Y.: Business Model Generation: A Handbook for Visionaries, Game Changers, and Challengers. John Wiley & Sons, Hoboken (2010)
34. Bondi, A.B.: Proceedings of the Second International Workshop on Software and Performance – WOSP 2000, p. 195 (2000)
35. Nielsen, C., Lund, M.: The Concept of Business Model Scalability (2015). https://doi.org/10.2139/ssrn.2575962
36. Boden, T.: The grid enterprise—structuring the agile business of the future. BT Technol. J. **22**(1), 107–117 (2004)
37. Szulanski, G., Jensen, R.J.: Growing through copying: the negative consequences of innovation on franchise network growth. Res. Policy **37**, 1732–1741 (2008)
38. Cramton, P.: Spectrum auction design. Rev. Ind. Organ. **42**(2), 161–190 (2013)
39. Hazlett, T.W., Muñoz, R.E.: A welfare analysis of spectrum allocation policies. Rand J. Econ. **40**(3), 424–454 (2009)
40. Matinmikko-Blue, M., Yrjola, S., Seppänen, V., Ahokangas, P., Hämmäinen, H., Latva-aho, M.: Analysis of spectrum valuation approaches: the viewpoint of local 5G networks in shared spectrum bands. In: IEEE International Symposium on Dynamic Spectrum Access Networks (DYSPAN), Seoul (2018)

41. FCC: Code of Federal Regulations (CFR), Title 47, Part 96. https://www.ecfr.gov/cgi-bin/text-idx?SID=d22828178369977efa1a34e5ca8f71c9&mc=true&node=pt47.5.96&rgn=div5. Accessed 20 Aug 2020
42. FCC: Code of Federal Regulations (CFR), Title 47, Part 15, Subpart H. https://www.ecfr.gov/cgi-bin/text-idx?SID=d22828178369977efa1a34e5ca8f71c9&mc=true&node=pt47.5.96&rgn=div5. Accessed 20 Aug 2020
43. ECC: Report 205 (2014)
44. 3rd Generation Partnership Project; Technical Specification Group Services and System Aspects; Study on Communication for Automation in Vertical Domains (Release 16), 3GPP TR 22.804, V16.1.0 (2018)
45. Ojanen, P., Yrjölä, S., Matinmikko-Blue, M.: Assessing the feasibility of the spectrum sharing concepts for private industrial networks operating above 5 GHz. In: 14th European Conference on Antennas and Propagation (EuCAP) (2020)
46. European Commission: Guide to the Radio Equipment Directive 2014/53/EU (2018)
47. Radio Equipment Directive 2014/53/EU of the European Parliament and of the Council (2014)
48. Ofcom: Shared access license. https://www.ofcom.org.uk/manage-your-licence/radiocommunication-licences/shared-access. Accessed 20 Aug 2020
49. FCC: Code of Federal Regulations (CFR), Title 47 Telecommunication. https://www.ecfr.gov/cgi-bin/text-idx?gp=&SID=e5e9b61ebcf975abfab699712a19750f&mc=true&tpl=/ecfrbrowse/Title47/47tab_02.tpl. Accessed 20 Aug 2020
50. FCC: Code of Federal Regulations (CFR), Title 47, Part 2, Subpart J. https://www.ecfr.gov/cgi-bin/text-idx?SID=273e14cf6e6820401103c2732ba484e0&mc=true&node=pt47.1.2&rgn=div5#_top. Accessed 20 Aug 2020
51. ACMA: Know what you must do, https://www.acma.gov.au/know-what-you-must-do. Accessed 18 Aug 2020
52. SRRC: SRRC Type Approval. http://www.srrccn.org/. Accessed 20 Aug 2020
53. OFCA I 401(19), Issue 27 (2019)
54. TEC: Mandatory Testing and Certification of Telecom Equipments (MTCTE). https://www.tec.gov.in/mandatory-testing-and-certification-of-telecom-equipments-mtcte%20/. Accessed 18 Aug 2020
55. MIC: Conformity Certification System. https://www.tele.soumu.go.jp/e/equ/index.htm. Accessed 20 Aug 2020
56. ICASA: Type Approval. https://www.icasa.org.za/pages/type-approval. Accessed 18 Aug 2020
57. ISED: RSP-100 — Certification of Radio Apparatus and Broadcasting Equipment, Issue 12 (2019)
58. IFT: Homologación. http://www.ift.org.mx/industria/homologacion. Accessed 18 Aug 2020
59. ENACOM: Inscripción de Equipos, https://www.enacom.gob.ar/inscripcion-de-equipos_p3076. Accessed 18 Aug 2020
60. ITU-R: Radio Regulations, vol. 1: Articles (2020)

Novel Spectrum Administration and Management Approaches Transform 5G Towards Open Ecosystemic Business Models

Seppo Yrjölä[1,2(✉)] and Pekka Ojanen[3]

[1] Nokia, Oulu, Finland
Seppo.yrjola@nokia.com, Seppo.yrjola@oulu.fi
[2] University of Oulu, Oulu, Finland
[3] Co-Worker Technology Finland Oy, Turku, Finland

Abstract. The ongoing 5G evolution transforming network from connectivity driven to service dominant logic will impact the stakeholder roles, ecosystem and business models. Systemic change will lower the barriers to entry and expand the ecosystem to new roles such as local operators, edge cloud services providers and resource aggregators and agents. Spectrum regulation has traditionally acted as a gate keeper of the mobile service provisioning, and lately national authorities have reacted via allocating new frequency bands and considering novel flexible spectrum administration and management methods and tools. This paper provides a comprehensive overview of the most recent spectrum regulation decisions for mobile communication networks and shows how local licensing, spectrum sharing, and unlicensed commons approaches work as novel business model antecedent. The study analyzes key spectrum antecedents for the open ecosystemic business model value configuration.

Keywords: Business model · Resource configuration · Spectrum administration · Spectrum management · 5G

1 Introduction

The fifth generation (5G) mobile communication network evolution is expanding and extending network services from mobile broadband to towards various industries with radically improved speed, capacity, time sensitiveness in connecting humans, machines and intelligence [1]. Transformation from present-day mobile network operator centric connectivity driven business [2] towards service dominant logic will be stemming from system architectural change leveraging softwarization, virtualization, network slicing, native-cloudification, novel spectrum management and service based architecture [3].

Wireless communication industry is prospecting novel and differentiating opportunities and value creation mechanisms to cope with the change and move beyond connectivity [1]. Access to human and industrial data has become central in value creation and advantage exploitation [4]. Mobile communication network and services related

G. Caso et al. (Eds.): CrownCom 2020, LNICST 374, pp. 158–175, 2021.
https://doi.org/10.1007/978-3-030-73423-7_12

business research have concentrated on mainstream communication service providers techno-economics founded on the mobile broadband and its service variations [2]. Transition from integrator to collaborative value configuration was discussed in [5] and novel stakeholder roles such as integrator, local-operator, neutral-host and market agent discussed in [6–9]. Key technology enablers and their impacts to business were analyzed in relation to cloudification [10], web-based service models [11], and Internet of things enablement [12].

Higher mmWave frequencies and ever-increasing variety of spectrum bands with local distinct industrial use cases have further fragmented spectrum regulation [13]. Nationwide long-term spectrum assignments are complemented by local licenses [14], dynamic and shared spectrum access [15–17] and unlicensed access [18] that represents a major paradigm change in spectrum management for mobile communication networks. The US citizens broadband radio service [16] and Europe originated licensed shared access (LSA) [17] managed spectrum sharing concepts were found to enable scalability and to extend business model design towards internet business models [19]. The valuations [13] and pricing [14] of private LTE and 5G local and shared spectrum were found as essential enablers in spectrum regulation.

This interdisciplinary study seeks to cover both the technology and business perspectives. Technology view stemming from product platform with components and interfaces is focusing modularity [20] exploring economies of scale. On the other hand, business research explores sustainable growth, creation of scalable ecosystems and replicability on novel markets [21] via innovatively matching needs and resources. In a contemporary research [22] platform definition is expanded to data and algorithms and found as enabler in transforming towards network-of-services model builds. This paper seeks to answer the research question: *How could the Novel Spectrum Administration and Management Approaches Transform 5G Towards Open Ecosystemic Business Models?*

The study provides a comprehensive overview of the most recent spectrum regulation decisions for mobile communication networks and shows how local licensing, spectrum sharing, and unlicensed commons approaches work as novel business model antecedent. Furthermore, the key spectrum antecedents for the open ecosystemic business model value configuration are analyzed.

The paper is organized as follows: after the introduction, the theoretical foundation is discussed. Third section provides a comprehensive overview of the most recent spectrum regulation and analyses enabled business model scenarios. Section 4 summarizes key findings and gives suggestions for future research.

2 Theoretical Foundation

2.1 Business Model Value Configuration

The value creation, value sharing, value delivery and value capture processes and related business model concept have become essential tools in business studies [23]. Business model framework has been traditional discussed from the action angle [24]. Contemporary research [25] views business models more from value creation, opportunity exploration, and competitive advantage exploitation angles [23, 26–28]. Ecosystem driven collaboration is driven by co-creation of opportunities and advantages [29] stemming from creation, delivery, sharing and capture of value [30]. The functioning and prosperous business models have been found to be scalable [31], replicable [32], and sustainable from business, social and environmental perspectives [33]. Growth potential builds on dynamic capabilities to scale internally and reiterate externally. The analysis of business model elements carried out in this research builds on above discussed elements as summarized in Fig. 1.

In addition to business model antecedents, the paper assess the impact of spectrum administration and management concepts from open value configuration and ecosystemic business model configurations perspectives [34]. We extend the *supply* and value chain focused focal firm centralized conceptualization [35–37] towards *demand* focused co-creation with customers, and finally considering novel *ecosystemic* model extending further to value co-capturing within the ecosystem [38]. In this paper we consider ecosystem to consists of the governance of network, platform keystones, complementors, open interfaces, innovative capabilities and resources, and modularity aspects [39].

2.2 Spectrum Management Archetypes

Spectrum management aiming at efficient utilization of the scarce national resource can be divided into three archetypes [13]: market-based mechanisms [40, 41], administrative assignment [42] and spectrum commons approach [43].

Market-Based Mechanism
The market-based mechanism [40] allows the markets to define who values the spectrum the most and should be granted the rights to use the spectrum in contrast to regulators. The most widely used mechanism is spectrum auction [44] that was the major method of assigning the 3G and 4G spectrum. Mechanism continued to limited number of individual spectrum access rights while the number of MNOs wishing to enter the market kept increasing. Furthermore, market-based approaches introduced flexibility into the market through spectrum trading and leasing options, which to date has not been taken into use widely [45].

Administrative Assignment
The traditional spectrum management approach has been administrative allocation where the national regulatory authorities (NRAs) have defined individual spectrum access rights, decided on the related rules and conditions, such as mandatory coverage obligations and assigned spectrum to cellular networks. The first generations of mobile connectivity market were in the hands of the regulators who solely command and control the number of licensees and decided on the admittance of any new entry. In the course of time, NRAs opened the mobile communication market for competition with state-owned monopolies that resulted in several benefits [46]. The administrative allocation faced growing criticisms on the fairness as continues to be deployed for 3G and 4G spectrum assignments in some countries [47]. Lately novel administrative methods have been revisited through regional and local licensing approaches [18].

Spectrum Commons

The unlicensed commons approach differs from the administrative allocation and market-based mechanisms as it allows market entry to a variety of wireless systems potentially deployed by a variety of stakeholders based on shared access to the spectrum instead of individual spectrum access rights [43, 48]. Traditionally, the unlicensed approach has not attracted mobile operators whose deployment was based on individual access rights providing exclusive access. On the other hand, in the 4G and 5G era to cope with exponentially growing data traffic, the cellular community has introduced 4G and lately 5G technology variants for unlicensed access in certain bands to share the bands with the other users [18]. The spectrum commons approach has been the source of success for wireless local area networks deployed by any stakeholder.

3 The Business Perspective of Spectrum Administration and Management Enablers

This section discusses the novel spectrum administration and management methods as enablers for the business models transformation. 5G spectrum management frameworks are already in place or are being defined for making dedicated spectrum available for wideband public and private networks, the main use case being MBB. The focus is on harmonized spectrum bands identified to international mobile telecommunications (IMT) in the Radio Regulations of the International Telecommunications Union (ITU) and covered by standards of the Third Generation Partnership Project (3GPP). Such bands are globally the major licensed bands for mobile communications. Suitable license exempt bands are also widely available, and some shared spectrum has been released and is being considered. The selected exemplary cases in this paper present spectrum management approaches in several countries on 1800 MHz, 2.3 GHz, 2.5 GHz, 3.5 GHz, 4 GHz, and 26/28 GHz licensed frequency bands considered for LTE and 5G. The license exempt usage of the 5 GHz Radio Local Area Network (RLAN) bands, 6 GHz, and 60 GHz is also addressed. Information on studied spectrum management frameworks have been collected both from public national regulatory authority (NRA) sources and interviews. The spectrum administration and management frameworks are analyzed based on assessment framework [18] developed for the novel industrial use cases and summarized in (Table 1).

3.1 Market Based Mechanism

Nationwide Individual Authorizations

The dominant approach used worldwide for awarding licenses for 5G public mobile networks is auctioning the individual authorizations where the associated nationwide spectrum assignments are awarded exclusively to the highest bidders. This approach has been and is used by most countries for authorizing most of the 4G and 5G networks and assigning the associated spectrum [49, 50]. The market based national licensing supports the traditional MNO-centric closed business model.

The market based approach has been used for authorizing nationwide deployments in 5G bands in countries like Denmark (700 MHz), Finland (3.4–3.8 GHz), France (700 MHz), Germany (3.4–3.7 GHz), Hungary (3.4–3.6 GHz), Ireland (3.4–3.6 GHz, 26 GHz), Italy (700 MHz, 3.6–3.8 GHz, 26 GHz), Spain (3.6–3.8 GHz), Sweden (700 MHz), UK (3.4–3.6 GHz), Canada (600 MHz), US (600 MHz, 24 GHz, 28 GHz, 37 GHz, 39 GHz, 47 GHz), Australia (3.6 GHz), Hong Kong (3.3 GHz, 3.5 GHz, 4.9 GHz), Korea (3.5 GHz, 28 GHz). The auctions are also planned to be used for 5G band authorizations e.g., in Austria (700 MHz), Belgium (3.6–3.8 GHz, 26 GHz, 32 GHz, 40 GHz), Finland (26 GHz), France (3.4–3.8 GHz, 26 GHz), Poland (700 MHz, 3.6–3.8 GHz) Spain (700 MHz, 26 GHz), Sweden (3.5–3.7 GHz), UK (700 MHz, 3.6–3.8 GHz), US (38 GHz, 47 GHz), Canada (3.5 GHz). In most cases the auctions are used for awarding nationwide licenses with exclusive usage rights, but they have also been used without full exclusivity. For example, in Italy, the 26 GHz band was auctioned employing an innovative Club Use licensing regulatory framework.

The 26 GHz in Italy (ITA 26 GHz)

The Italian regulator AGCOM auctioned the 26 GHz band as part of a 5G multi-band auction [51]. All five lots of 200 MHz from the 26.5–27.5 GHz were awarded to the five highest bidders, one lot for each [52]. The 26 GHz licenses were awarded on a nationwide basis without coverage obligation and fully exclusive usage rights. Instead, the licenses are based on a so-called *"Club Use"* model, in which each licensee can use the available spectrum in areas where the other licensees do not use it. Once the actual licensee wants to deploy its networks in the area, the individual rights of use on the acquired spectrum will prevail. This mechanism is intended to increase the efficiency of the overall use of the 26 GHz band. Due to the propagation characteristics of the band, the coverage areas are expected to be small, and therefore the regulation allows wide freedom for the licensees to agree on building the coverage in collaboration. There was low interest towards the 26 GHz auction, probably at least partly due to the Club-use model, as the MNOs prefer dedicated exclusive bands.

Regional/Local Individual Authorizations

It is also possible to use market based mechanism for awarding the regional/local individual authorizations and the associated spectrum assignments to the highest bidders. Due to regional/local coverage areas this approach leads to geographical sharing, which may lead to higher overall spectrum use efficiency. But depending on the sizes of the areas and the size of the country, the number of auctioned regional licenses may be large, and yet the areas larger than required for actual network deployments.

The 3.5 GHz Citizens Broadband Radio Service for PAL use (US CBRS PAL)

Citizens Broadband Radio Service (CBRS) system made 3.55–3.7 GHz band available in the US [16]. The band and its users consists of three layers: incumbents users (tier 1), individually authorized priority access license users in 3.55–3.65 GHz (PAL, tier 2) and general authorized "license by the rule" access users (GAA, tier 3) in the whole band.

In the PAL spectrum auctions the band was divided into 10 MHz channels. Maximum PAL spectrum holding is four times 10 MHz in any service area for 10 years with a renewal option. The PAL authorizations are required to ensure interference protection to the incumbents. The novelty of PAL regulation is that it allows the leasing of unused spectrum. Moreover, general authorized access (GAA) users are allowed to access the unutilized incumbent or PAL licensed spectrum slots. The spectrum management is based on the spectrum access system (SAS) that grants available spectrum for the base stations (CBSDs) while ensuring protection to higher tier users. The scalability and flexibility of the CBRS system enables new entrants and business models: PAL licenses allowing for acquiring exclusive access, GAA offering a low cost option for non-critical services and additional option between the PAL and GAA. The leasing rules for CBRS PAL spectrum provide also a potentially lower cost option for industrial enterprises to lease exclusive spectrum at their facility utilizing the SAS provided spectrum marketplace for PAL spectrum. In August 2020, the PAL auctions bidding concluded and raised a total of $ 4,5 Billion in net bids, with 228 bidders winning a total of 20,625 licenses. Top winners included Verizon, Dish, and several the top US cable companies. The auction was unique in a way that many qualified bidders were non-traditional auction participants thanks in part to the smaller size of the licenses. Utilities, rural service providers, universities and others joined wireless and cable service providers in the bidding.

3.4–3.8 GHz *in Austria (AUT* 3 GHz)
The bands 3.41–3.6 GHz and 3.6–3.8 GHz were auctioned in early 2019 [53]. The country was divided into 6 urban and 6 rural regions, so that there was one urban region covering one or two of the major cities inside each rural region. Technical conditions were those defined by the EU. The auction resulted in three MNO's getting access to 100–140 MHz of spectrum each, over all regions, thus building up nationwide coverage, while the remaining spectrum was all sold to other organizations in 5 out of 12 regions. However, a portion of the spectrum remained unsold in 7 regions. In two regions the unsold amount was 10 MHz, in one region 40 MHz, and in four regions 60 MHz in each. Therefore, the regional auctions did not result in efficient overall use of the band over the whole country. One reason may be that the cost of spectrum even in the rural regions ended up high, e.g., the cost of 30 MHz sold in one rural region was 1.8 million €, and the cost of 40 MHz in another rural region was 4.3 million € [54].

3.2 Administrative Assignment

Nationwide Individual Authorizations Spectrum was traditionally assigned to MNOs through administrative assignment, in so called beauty contest. The currently dominant method is market-based approach through auctions, but the administrative assignment is still used in a few countries, such as Japan and China for awarding exclusive nationwide spectrum assignments. In April 2019, The Japanese Ministry of Internal Affairs and Communications (MIC) assigned spectrum in the 3.7, 4.0, 4.5 and 27/29 GHz bands through a beauty contest to four mobile operators, assigning dedicated sub-bands for each. This approach can also be used to assign shared spectrum. China has recently awarded the band 3.3–3.4 GHz to three mobile network operators on a shared basis. It is the first time China makes IMT spectrum available for shared use. The band is available nationwide, but only for indoor use, which makes sharing straightforward.

Regional/Local Individual Authorizations

The dominant authorization procedure for issuing regional/local licenses is administrative assignment. The most common approach is first-come first-served, used especially in cases where the number of licenses in a certain area/location is limited by co-existence requirements. In other cases, the approach can be all-come all-served, when the number of licenses is not limited. Usually there is a yearly fee for the spectrum usage, either a fixed amount, or depending on coverage, bandwidth and possibly some other parameters. Some frameworks offer protection from harmful interference between the regionally/locally deployed networks based on coordination by the NRA or by technical means, while in other cases the approach is uncoordinated, leaving avoidance of harmful interference to the licensees, or to be covered by technical requirements. By employing this regulatory framework, the NRAs intend to respond to the foreseen local and regional spectrum requirements of the verticals.

Shared Access Mobile Bands in the UK (UK Shared)

The UK regulator Ofcom has made spectrum in four shared access bands locally available through Shared Access licenses. The bands are 1.8 GHz, 2.3 GHz, 3.8–4.2 GHz and the 26 GHz [55]. The access is individually authorized, allowing operation in a certain location, for some of the bands only indoors. The applicants need to specify the bands they would like to access, as well as the planned locations. There are two types of licenses, low power license (per area license) and medium power license. The low power license allows the users to deploy a required number of base stations in a circular area of a 50 m radius, while the terminals are covered by the same license. The medium power licenses are available mainly for deployments in the rural areas. The licenses are assigned on a first-come first-served basis and the access is coordinated by Ofcom to ensure avoidance of harmful interference between the users. This approach can provide certainty for the spectrum access and a possibility to provide QoS. The yearly license fees are cost based administrative fees, reflecting Ofcom's cost of issuing the license. The licenses are valid for an indefinite duration.

Local Access to Unused Spectrum of the MNOs in the UK (UK Local)

The Local Access license provides a way for users to access spectrum licensed to MNOs in locations where the MNO is not using its full spectrum [56]. All bands assigned to MNOs are candidate bands. The Ofcom will issue a Local Access license if the new usage does not cause harmful interference to the MNO, and the MNO does not raise justified objections. The technical conditions will be considered on a case by case basis. The default license period is 3 years, and there is a one-off license fee of £950. There is a possibility for longer term or shorter term license. There is a lively 3GPP technology ecosystem for all MNO bands and equipment widely available, but the license will also allow deployment of other technologies.

The 2.3 GHz in Finland (FIN 2.3 GHz)

The 2.3 GHz band is identified globally for IMT and the European regulation for mobile networks is in place. However, in Finland and many European countries the band is used for other services than MBB, and re-farming would be impractical. The LSA concept developed by the European conference of postal and telecommunications administrations

(CEPT) could facilitate shared use. The regulation and standards exist [17], but the public mobile network operators have not shown interest towards accessing the band. As there are unused spectrum resources, the sub-band 2.3–2.32 GHz is being allocated in Finland to mobile service on a secondary basis and designated to private mobile networks, such as private LTE networks [57]. The specific 20 MHz band may provide the required spectrum for verticals, but its usability still depends on the geographical locations of the incumbent usage, i.e. wireless cameras that need to be protected from harmful interference. The 26 GHz band will be made available in Finland for nationwide 5G deployments as three 800 MHz bands. The sub-band 24.25–25.1 GHz is not to be auctioned and will be reserved for local private and industrial 5G deployments.

The 2.6 GHz in France (FRA 2.6 GHz)
The band 2.575–2.615 GHz was made available for private networks in 2019 [58]. The regulator's data base based web page shares the availability of frequencies to facilitate the application process. Furthermore, the applications are made available for the public reviewing. The local licenses have a maximum bandwidth of 40 MHz and are granted for 10 years. The yearly fee is 17 k€ per 5 MHz for the 100 km^2 coverage area. The relatively high fee may limit the possibilities of deploying some small scale innovative applications over a small area. The available bandwidth is likely to limit the number of networks in the same location. France has opened the 26 GHz band for 5G trials platforms, authorized for the period of three years. Some of the trials are led by verticals and those trials are focusing on industrial/private applications. The ACREP will determine the regulation for the band after the 3 year trials and based on the nature of the trials, regulation allowing for local deployments can be expected.

The 3.6 GHz Wireless Broadband Service in Canada (CAN WBS)
Wireless broadband service (WBS) band 3.65–3.7 GHz is allocated to fixed and mobile systems for Tier 4 service areas across Canada [59]. The service area specific annual fee for a "all-come-all-served" license depend on the population density, ranging from 250 CAD in rural areas up to 21 kCAD in Toronto. The licensees share the band within the service area and are obligated to self-manage the coexistence facilitated by the regulator hosted spectrum management system (SMS) database showing the status of licenses and related radio characteristics. WBS spectrum is easily accessible, but depending on the location, the coordination that is left to the licensees may be an extra burden. ISED studies the need for reorganizing the band and updating the regulation, but no short-term changes are expected as the WBS band is widely used.

The 3.5 and 3.7 GHz in the Netherlands (NL 3 GHz)
The 3.41–3.5 GHz and 3.7–3.8 GHz bands are used by a military satellite earth station in the northern part of the country. In order to protect the incumbent usage, those frequencies cannot be used for mobile services in the northern part of the country. Outside of the above discussed restriction zones, base station specific 40 MHz bandwidth licensed are available for local private networks, under specific operational regulation [60]. To date, more than 150 licenses have been issued and in some areas the bands are getting fully occupied. As the authorization process is based on the first-come first-served principle, in most popular areas it may not be possible to get a license. For the moment, the licenses

are temporary, as there are plans to remove the military earth station from the band and reorganize the use of the whole C-band around 2022. It seems that one option under consideration would be to re-farm the local networks to the 3700–3800 MHz band as done in Germany and planned in Sweden. The C-band is one of the European pioneer bands for 5G, and the reorganization of the band could release most of the C-band available for public 5G networks in an efficient manner.

The **3.7 GHz** *in Germany (GER* **3.7 GHz***)*

The band 3.4–3.7 GHz was auctioned in 2019 for public mobile networks whereas the band 3.7–3.8 GHz was made available for individually authorized local private assignments [61]. Applications can be submitted any time; eligibility is related to the land ownership or right of use. The license duration is 10 years, and the licenses are transferable. There is a fee depending on the assignment bandwidth, license duration and the category of the deployment area [62]. Deployment in densely populated area is more expensive than a deployment in a sparsely populated area. The locations and the area can be defined by the applicants. The approach is service and technology neutral. Efficient use of spectrum is required, with a principle use-it-or-lose-it. There are technical requirements to ensure that no harmful out-of-band interference is created. In addition, operators of geographically adjacent networks are obliged to negotiate agreements between them. If this fails, the regulator BNetzA may define measures to ensure efficient and interference free use of spectrum for all affected operators. This could include definition of a maximum field strength limit at the edge of the coverage area. 74 licenses have been awarded by BNetzA by September 2020. Furthermore, there are also around 50 experimental assignments.

The BNetzA assumes that also the 26 GHz band could be used by various local 5G applications [63]. The main characteristics of the proposed regulatory framework are similar to the regulation of the 3.7 GHz band. Usage of general authorization was not felt possible due to demanding requirements for incumbent protection in the band. The 26 GHz band could complement the 3.7 GHz band by providing much wider bandwidths for shorter range communications required by many industrial applications.

The **3.8–4.2 GHz** *in Belgium (BEL* **4 GHz***)*

The Belgian regulator BIPT intends to allow 4G and 5G private networks in the 3.8–4.2 GHz band [64]. The licenses will be local, and not transferable. The maximum amount per licensee is limited to 40 MHz, which is to facilitate sufficient overall capacity for multiple licensees in one location/area. A local network license is based on an applicant defined circular zone with minimum radius of 100 meters, allowing base stations inside the zone. Terminals are allowed outside the circular area as long as connected to one of the base stations that inside the licensed circular zone. The BIPT makes a compatibility study for the applications, determines the technical conditions and assigns the frequencies. There is an initial fee for each new application and annual fees per area, to facilitate high number of small cells inside the licensed area. The BIPT assumes this to become a typical 5G deployment scenario. The licenses are expected to become available in 2021.

The 28 GHz *Shared Spectrum in Hong Kong (HK 28 GHz)*

In the 27.95–28.35 GHz band, four 100 MHz channels are shared geographically for the local wireless broadband services and assigned by the "first-come-first-served" rule. Real time voice communications to and from public network is not allowed. The license duration for the maximum 400 MHz bandwidth within the maximum area of 50 km^2 is 5 years with 5 year extension option [65]. In the concept, the potential co-existence issues are self-managed by the peer licensees. The regulatory authority decides upon the amount of granted spectrum, based on the application. Annual spectrum license fee depends on the number of base stations, number of devices and allocated bandwidth.

The 28 GHz *in Japan (JPN 28 GHz)*

The Japanese regulator targets at enabling industrial automation across verticals via "Local 5G" spectrum initiative for non-existing mobile operators [66]. The 28.2–28.3 GHz was allocated first and will be followed by the band 28.3–29.1 GHz till end of 2020, depending on the sharing studies with the satellite incumbents. The licensing is based on the land ownership and in case of the land or property is not owned by the applicant, allowed service will be fixed wireless only. There is a fee for base stations and for terminals. The Kanto Bureau of Telecommunications granted Japan's first private 5G radio station provisional license in the band 28.2–28.3 GHz band in February 2020.

3.3 Spectrum Commons

The license exempt bands, spectrum commons, allow operation of compliant radios under a general authorization, without an individual authorization. Technical or regulatory means are employed to facilitate co-existence with other applications in the band.

The 3.5 GHz *CBRS for GAA Use in the Unites States (US CBRS GAA)*

The general authorized access (GAA) users can access the portions of the CBRS band 3.55–3.65 GHz that are unused by the incumbents and the PAL users and the portions of the band 3.65–3.7 GHz that are unused by the incumbents [16]. The GAA users can access the bands on an unlicensed, shared basis. The amount of GAA spectrum may vary based upon variations incumbent and PAL usage, and the unlicensed GAA users may experience harmful interference from higher level users and other GAA users. The spectrum access system (SAS) will identify suitable spectrum for the base stations (CBSDs). Operation under GAA on the CBRS band could be suitable for new entrants and certain business models, because GAA usage allows for a very low cost option for non-mission critical services.

The 5 GHz *Radio Local Area Network*

The ITU radiocommunication sector has defined the 5.15–5.35 GHz and 5.47–5.725 GHz for wireless access systems including radio LANs [67]. The most common authorization framework deployed globally is a general "unlicensed" authorization. Because several incumbents use those bands, a number of technical and operational requirements have been defined by the ITU-R and in the standardization to avoid interference to incumbents. European telecommunications standards institute standardization allows system deployment in the EU, in additional countries covered by the European conference of postal and telecommunications administrations regulation, and several additional countries outside of Europe, recognizing the EU CEPT regulation. Similarly, the unlicensed national information infrastructure (U-NII) radio band 5.725–5.850 GHz in US has been allocated and specified to unlicensed RLAN type of devices [68] and allocated by several national regulators globally for similar application. LTE unlicensed (LTE-U), licensed spectrum access (LAA) and Multefire are 3GPP 4th generation (4G LTE) based standards were developed to access the license exempt spectrum while being fully compliant with the RLAN standards and regulation. To complement the 3GPP 5G standardization, in the current release 16 the 3GPP has defined 5G new radio unlicensed (5NR-U). Depending on the regulatory framework, the so far low allowed maximum transmit power in the band 5.15–5.35 GHz may limit the coverage area to small cells.

The 6 GHz Radio Local Area Network

To meet the exponentially growing need for wireless "last mile" capacity and to cope with congested 2.4 GHz and 5 GHz RLAN bands, there are several studies and recent allocations on expanding the RLAN use to bands above 5 GHz. In the US, additional 5.925–7.125 GHz band, divided into four segments, has been allocated for licensed exempt devices with similar technical regulatory conditions than in the 5 GHz spectrum [69]. Two segments allow only low-power operations while other two 6 GHz segments require employment of automated frequency coordination (AFC) system to protect incumbent services via providing automated frequency availability information. European regulatory study on wireless access system is focusing on 5.925–6.425 GHz spectrum. Radio local area networks are allowed with limited transmit power indoor with strict emission requirements to protect large installed base [70]. The UK regulator Ofcom will make the band 5.925–6.425 GHz available for RLANs and other related wireless technologies on a license-exempt basis, enabling indoor use and very low power (VLP) outdoor use [71]. A regulatory challenge for large scale availability of the 6 GHz band may be that it is not identified for IMT by the ITU-R, and the technical regulatory requirements are likely to become different in different regulatory frameworks, but on the other hand the 3GPP has developed a 5G NR-U standard to the 6 GHz band as part of Release 16. The upper part of the band is on the agenda of the WRC-23.

The 26 GHz in Australia (AUS 26 GHz)

The Australian regulatory authority is in process of allocating 24.25–27.5 GHz band for fixed and mobile applications in 2020 [72]. Several authorization schemes are foreseen for different parts of the band: class license (license exempt), apparatus license and spectrum license. The 24.25–24.7 GHz segment will become available countrywide for private broadband indoors without interference protection. The 24.7–25.1 GHz spectrum will be licensed to indoor and outdoor wireless broadband using the apparatus licensing, limited to private property. Furthermore, the 25.1–27.5 GHz will be allocated to wide-area wireless broadband in 34 metropolitan areas and regional centers via auction. The wide range of authorization options can support deployment of innovative 5G applications.

The 60 GHz

Spectrum around 60 GHz has been widely available for several years for license exempt data networks. The exact bands depend on the country, but in general the 60 GHz band offers 5–14 GHz of spectrum for very high bitrate data, video and audio applications that supplement the capabilities of Wireless LAN devices. Regulation exists for operation in the band 57–71 GHz in the US [73] and in Europe in the band 57–64 as short range device (SRD) and in 57–71 GHz as wideband data transmission system [74]. The attractiveness of the band was further increased by the recent WRC-19, which identified the band 66–71 GHz globally for IMT. Currently, both IEEE and ETSI standards exist, and the 3GPP is in the process of preparing a NR-U standard for the band as part of its Release -18. The band is suitable for very high bitrate transmissions due to the up to several GHz bandwidths. The frequency range does not offer wall penetration, but free space ranges up to 300–500 m can be reached with highly directional antenna.

Table 1. The summary of the spectrum administration and management approaches.

Framework	Market based	Administrative	Commons
Auctioned 5G	Nationwide		
JPN 4 GHz	Nationwide		
ITA 26 GHz	Nationwide		
US CBRS	Regional (PAL)		Local (GAA)
AUT 3 GHz	Regional		
UK Shared		Local	
UK Local		Local	
FIN 2.3 GHz		Local	
FRA 2.6 GHz		Local	
CAN WBS		Regional, shared	
NL 3 GHz		Local	
GER 3.7 GHz		Local	
BEL 4 GHz		Local	
GER 26 GHz		Local	
FIN 26 GHz	Nationwide	Local	
HK 28 GHz		Local, shared	
JPN 28 GHz		Local	
RLAN 5 GHz			Nationwide, shared
RLAN 6 GHz			Nationwide, shared
AUS 26 GHz	Regional	Local	Local, shared
60 GHz			Nationwide, shared

3.4 Open Ecosystemic Business Antecedents

In 5G ecosystem, stakeholders have a wide variety of novel assets and resource as well as needs that should be orchestrated and configured optimally in order to co-create, share and co-capture value. In identifying, matching and bridging needs and resource companies can take multiple roles in the ecosystem [75]. Novel spectrum management approaches could assist the progression of value-creation processes and transformation from closed integrator and collaborator models towards ecosystem-focused transaction and bridging models.

In the *integrating* resource configuration, a company source all the resources utilizing traditional value-chain logic. This closed business model has been widely deployed by incumbent mobile network operators leveraging nation-wide licensed spectrum in offering mobile communication services.

In a *collaborator* model, a focal firm completes its offering and creates value via orchestrating and configurating partner's complements. Mobile broadband business has utilized this model in bundling connectivity services with media content and commerce-based mobile banking service offering as well as in the mobile virtual network operator (MVNO) model. Furthermore, emerging spectrum trading, leasing and partnering models enabled by majority of NRAs are emerging in offering dedicated services to verticals.

A *transactional* marketplace builds on a digital multi-sided-platform that extends and reduces friction to access resources through intelligent matching. From spectrum management perspective, the first phase NRA acting as a focal orchestrating firm for local licensing. In the recently commercially opened, US CBRS spectrum sharing model spectrum management and transactional platform is outsourced to SAS operators.

In bridge resource configuration a platform creates value via connecting the needs and resources of unconnected stakeholders without owning the virtualized resources. CBRS PAL auctions concluded in August 2020 will open new opportunities for a bridge provider via a SAS marketplace matching spectrum supply and demand.

Exploring and exploiting *opportunities* and *advantages* can be seen to motivate ecosystemic interaction from a dynamic capability perspective. Novel opportunities were found in utilization of local spectrum, unlicensed spectrum offering and shared spectrum marketplace in offering connectivity services to growing industrial automation segment. *Value* creation, delivery, sharing and capture are considered the key elements of a functioning business model. Transformation from tardy nation-wide multi-million spectrum licensing towards distinct local valuation and pricing, automated low transaction cost administration and sharing economy-based spectrum sharing are creating new value. As the mobile broadband business has started to even out, industrial automation across verticals is seen as a new value capture opportunity with higher willingness to pay based on quality and service level agreement. Timely access to affordable exclusive spectrum based on use case and business needs will lower entry barrier and create advantage particularly to novel stakeholders like micro-operators. As summary, enablers for 5G growth via *scalability* and *replicability* were found in radio standardization at 3GPP and IEEE, zero touch automation and transaction platforms. Furthermore, compared to traditional spectrum administration and management, novel local licensing and sharing were found essential contributors to spectrum resource *efficiency* and *sustainability*. In addition to discussed enablers, there are a few framing elements that would need to be considered in applying novel approaches into 5G systems. In the mobile operator business, network deployment details are considered business critical information that many novel database based spectrum administration and management concepts request. In additions to ever increasing variety of spectrum bands, fragmented regulation due to national policies and differentiating incumbent usage can seriously reduce scalability and replicability of the related technology platforms. Furthermore, co-existence and interference management between neighboring operators calls for new technologies. Figure 1 summarizes how 5G business exploiting novel spectrum management can be built on novel business opportunities, value generation and competitive advantage that have positive strategic consequences on scale, reproduction, and stability. Stemming from 5G system level value driver summarized in Fig. 1, spectrum administration and management enablers can be seen to transform business model value-configurations towards open and ecosystemic as depicted in Fig. 2. In addition to spectrum commons, both the administrative local licensing and CBRS concept were found to democratize the tools of production through access to affordable spectrum, cutting the costs of consumption by democratizing distribution with web-scale automatization and connecting supply and demand via NRA database and further utilizing automated SAS marketplace.

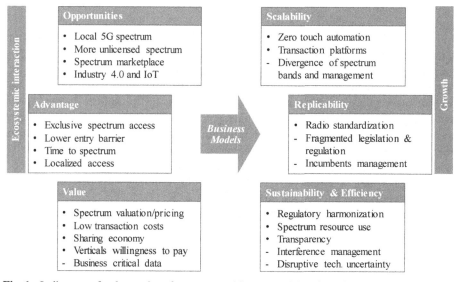

Fig. 1. Indicators of value and performance enablers (·) and framing elements (-) of the future 5G business exploiting spectrum management enablers.

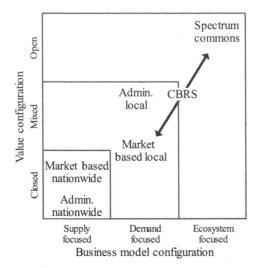

Fig. 2. Spectrum administration and management enablers in the evolutionary 5G business model [34].

4 Conclusions

This research explored the future of radio frequency administration and management utilizing ecosystem and business model frameworks. The ecosystemic open value configuration business model was found as key antecedent to enable wider cross-industry entry and involvement. scenario. The results indicate the importance of the transactional spectrum marketplace as a trigger for business model change. Local licensing and spectrum sharing methods and technologies have potential to transform the spectrum asset orchestrations and market. Theoretical implications of the study pave the way to utilize value and ecosystemic business model configuration in analyzing novel mobile business. In the paper the recent status of traditional spectrum administration and novel management approaches are reviewed through the lenses of evolutionary 5G business models focusing on ecosystem and growth antecedents.

Future research on the local licensing and database-based dynamic spectrum access approaches could consider stakeholder and ecosystem analysis. Further work is needed to validate novel spectrum administration and management technology and regulatory enablers in order to reduce concerns related to system reliability and trustworthiness.

Acknowledgment. The research has been supported by the Business Finland 5G Vertical Integrated Industry for Massive Automation (5G-VIIMA) program. The authors would like to acknowledge the support of the 6G Flagship programme at the University of Oulu.

References

1. Ahokangas, P., et al.: Business models for local 5G micro operators. IEEE TCCN **5**(3), 730–740 (2019)

2. Ahokangas, P., Matinmikko, M., Yrjölä, S., Okkonen, H., Casey, T.: "Simple rules" for mobile network operators' strategic choices in future cognitive spectrum sharing networks. IEEE Wirel. Commun. **20**(2), 20–26 (2013)
3. Yrjölä, S., Ahokangas, P., Matinmikko-Blue, M.: Novel context and platform driven business models via 5G networks. In: 2018 IEEE PIMRC, Genova, Italy (2018)
4. Latva-aho, M., Leppänen K. (eds.): 6G WP Key Drivers and Research Challenges for 6G Ubiquitous Wireless Intelligence. University of Oulu (2019). ISBN 978-952-62-2354-4
5. Noll, J., Chowdhury, M.M.R.: 5G: service continuity in heterogeneous environments. Wireless Pers. Commun. **57**(3), 413–429 (2011). https://doi.org/10.1007/s11277-010-0077-6
6. Ballon, P.: The platformisation of the European mobile industry. Commun. Strateg. **75**(15), 15–33 (2009)
7. Matinmikko, M., Latva-aho, M., Ahokangas, P., Yrjölä, S., Koivumäki, T.: Micro operators to boost local service delivery in 5G. Wireless Pers. Commun. **95**(1), 69–82 (2017). https://doi.org/10.1007/s11277-017-4427-5
8. Rasheed, T., et al.: Business models for cooperation. In: Radwan, A., Rodriguez, J. (eds.) Energy Efficient Smart Phones for 5G Networks. SCT, pp. 241–267. Springer, Cham (2015). https://doi.org/10.1007/978-3-319-10314-3_9
9. Valtanen, K., Backman, J., Yrjola, S.: Blockchain powered value creation in the 5G and smart grid use cases. IEEE Access **7**(1), 25690–25707 (2019)
10. Zhang, N., Cheng, N., Gamage, A.T., Zhang, K., Mark, J.W., Shen, X.: Cloud assisted HetNets toward 5G wireless networks. IEEE Commun. Mag. **53**(6), 59–65 (2015)
11. Gonçalves, V., Ballon, P.: Adding value to the network: mobile operators' experiments with Software-as-a-Service and Platform-as-a-Service models. Telematics Inform. **28**(1), 12–21 (2011)
12. Sarfaraz, A., Hämmäinen, H.: 5G transformation: how mobile network operators are preparing for transformation to 5G? In: 2017 Internet of Things Business Models, Users, and Networks, Copenhagen, pp. 1–9 (2017)
13. Matinmikko-Blue, M., Yrjölä, S., Seppänen, V., Ahokangas, P, Hämmäinen, H., Latva-aho, M.: Policy analysis of spectrum valuation elements for local 5G networks: case study of 3.5 GHz band. IEEE Trans. Cogn. Commun. Netw. **5**(3), 741–753 (2019)
14. Kokkinen, H., Yrjölä, S., Engelberg, J., Kokkinen, T.: Pricing private LTE and 5G radio licenses on 3.5 GHz. In: Moerman, I., Marquez-Barja, J., Shahid, A., Liu, W., Giannoulis, S., Jiao, X. (eds.) CROWNCOM 2018. LNICST, vol. 261, pp. 133–142. Springer, Cham (2019). https://doi.org/10.1007/978-3-030-05490-8_13
15. FCC: Second Memorandum Opinion and Order in ET Docket Nos. 02-380 and 04-186. 25 FCC Rcd 18661, vol. 23 (2010)
16. FCC: Part 96 - Citizens Broadband Radio Service (2015)
17. ECC: Report 205 Licensed Shared Access (LSA) (2014)
18. Ojanen, P., Yrjölä, S., Matinmikko-Blue, M.: Assessing the feasibility of the spectrum sharing concepts for private industrial networks operating above 5 GHz. In: EuCAP 2020, Copenhagen, Denmark (2020)
19. Yrjölä, S., Ahokangas, P., Matinmikko, M.: Evaluation of recent spectrum sharing concepts from business model scalability point of view. In: 2015 IEEE International Symposium on Dynamic Spectrum Access Networks (DySPAN), Stockholm, pp. 241–250 (2015)
20. de Mattos, C.S., Fettermann, D.C., Cauchick-Miguel, P.A.: Service modularity: literature overview of concepts, effects, enablers, and methods. Serv. Ind. J. (2019). https://doi.org/10.1080/02642069.2019.1572117
21. Gawer, A.: Bridging differing perspectives on technological platforms: toward an integrative framework. Res. Policy **43**(7), 1239–1249 (2014)

22. Yrjölä, S., Ahokangas, P., Matinmikko-Blue, M.: Novel platform-based ecosystemic business models in the future mobile operator business. Short Paper Presented at the 3rd Business Model Conference, New York (2019)
23. Zott, C., Amit, R., Massa, L.: The business model: recent developments and future research. J. Manag. **37**(4), 1019–1042 (2011)
24. Onetti, A., Zucchella, A., Jones, M.V., McDougall-Covin, P.P.: Internationalization, innovation and entrepreneurship: business models for new technology-based firms. J. Manage. Governance **16**(3), 337–368 (2012). https://doi.org/10.1007/s10997-010-9154-1
25. Amit, R., Zott, C.: Value creation in E-business. Strateg. Manag. J. **22**(6–7), 493–520 (2001)
26. Morris, M., Schindehutte, M., Allen, J.: The entrepreneur's business model: toward a unified perspective. J. Bus. Res. **58**(6), 726–735 (2005)
27. Teece, D.: Business models, business strategy and innovation. Long Range Plan. **43**(2–3), 172–194 (2010)
28. McGrath, R.: Business models: a discovery driven approach. Long Range Plan. **43**(2–3), 247–261 (2010)
29. Gomes, J.F., Iivari, M., Pikkarainen, M., Ahokangas, P.: Business models as enablers of ecosystemic interaction: a dynamic capability perspective. Int. J. Soc. Ecol. Sustain. Dev. **9**(3), 1–13 (2018)
30. Osterwalder, A., Pigneur, Y.: Business Model Generation: A Handbook for Visionaries, Game Changers, and Challengers. Wiley, New Jersey (2010)
31. Stampfl, G., Prügl, R., Osterloh, V.: An explorative model of business model scalability. Int. J. Prod. Dev. **18**(3–4), 226–248 (2013)
32. Aspara, J., Hietanen, J., Tikkanen, H.: Business model innovation vs replication: financial performance implications of strategic emphases. J. Strateg. Mark. **18**(1), 39–56 (2010)
33. Schaltegger, S., Hansen, E., Lüdeke-Freund, F.: Business models for sustainability: origins, present research, and future avenues. Organ. Environ. **29**(1), 3–10 (2016)
34. Xu, Y.: Open business models for future smart energy: a value perspective. Acta Universitatis Ouluensis. G, Oeconomica, University of Oulu, Oulu (2019)
35. Porter, M.E.: What is strategy? Harvard Bus. Rev. **Nov-Dec 1996**, 61–78 (1996)
36. Massa, L., Tucci, C., Afuah, A.: A critical assessment of business model research. Acad. Manag. Ann. **11**(1), 73–104 (2017)
37. Casadesus-Masanell, R., Ricart, J.E.: How to design a winning business model. Harvard Bus. Rev. **89**(1/2), 100–107 (2011)
38. Bereznoi, A.: Business model innovation in corporate competitive strategy. Probl. Econ. Transit. **57**(8), 14–33 (2015)
39. Mazhelis, A., Mazhelis, O.: Software business. In: ICSOB 2012, Cambridge, MA, USA, pp. 261–266 (2012)
40. Melody, W.H.: Radio spectrum allocation: role of the market. Am. Econ. Rev. **70**(2), 393–397 (1980)
41. Beltran, F.: Accelerating the introduction of spectrum sharing using market-based mechanisms. IEEE Commun. Stand. Mag. **1**(3), 66–72 (2017)
42. Levin, H.J.: Spectrum allocation without market. Am. Econ. Rev. **60**(2), 209–218 (1970)
43. Carter, K.R.: Policy lessons from personal communications services: licensed vs. unlicensed spectrum access. CommLaw Conspectus **15**(1), 93–117 (2006)
44. Cramton, P.: Spectrum auction design. Rev. Ind. Organ. **42**(2), 161–190 (2013). https://doi.org/10.1007/s11151-013-9376-x
45. Valletti, T.M.: Spectrum trading. Telecommun. Policy **25**(10–11), 655–670 (2001)
46. Anker, P.: From spectrum management to spectrum governance. Telecommun. Policy **41**(5–6), 486–497 (2017)
47. Kuroda, T., Forero, M.: The effects of spectrum allocation mechanisms on market outcomes: auctions vs beauty contests. Telecommun. Policy **41**(5–6), 341–354 (2017)

48. Bazelon, C.: Licensed or unlicensed: the economic considerations in incremental spectrum allocations. IEEE Commun. Mag. **47**(3), 110–116 (2009)
49. Analysys Mason: International comparison: licensed, unlicensed, and shared spectrum 2017–2020 (2020)
50. European 5G Observatory: Quarterly Report 6 (2019)
51. AGCOM: Delibera N. 231/18/CONS (2018)
52. Petracca, M.: The Agcom's regulation for the award and use of the 700 MHz, 3.6–3.8 GHz and 26.5–27.5 GHz bands to foster the transition to 5G technology. In: Joint EMERG-BEREC Workshop (2019)
53. RTR: Procedure for Spectrum Award in the 3410 to 3800 MHz Range (Non-binding translation), Tender Document (2018)
54. RTR homepage. https://www.rtr.at/en/tk/5G-Auction-Outcome. Accessed 14 Feb 2020
55. Ofcom: Shared Access license, Guidance document (2019)
56. Ofcom: Local Access License, Guidance document (2019)
57. Traficom: Public consultation. https://www.lausuntopalvelu.fi/FI/Proposal/Participation?proposalId=3bf8ce50-287d-476c-a0ac-44211d32f6e0. Accessed 25 Feb 2020
58. ARCEP Homepage. https://www.arcep.fr/demarches-et-services/professionnels/transformation-numerique-des-entreprises/guichet-frequences-2-6-tdd.html. Accessed 25 Feb 2020
59. ISED: SRSP-303.65, Technical Requirements for Wireless Broadband Services (WBS) in the Band 3650–3700 MHz (2010)
60. Agentschap Telecom wb page. https://www.agentschaptelecom.nl/onderwerpen/internetverbinding-verbeteren. Accessed 26 Feb 2020
61. BNetzA: Verwaltungsvorschrift für Frequenzzuteilungen für lokale Frequenznutzungen im Frequenzbereich 3.700–3.800 MHz (VV Lokales Breitband) (2019)
62. BNetzA: 5G spectrum fees for local usages, press release (2019)
63. BNetzA: Entwurf der grundlegenden Rahmenbedingungen für 5G Anwendungen im Bereich 26 GHz (24,25 - 27,5 GHz), consultation (2019)
64. BIPT: Consultation at the request of the Minister of Telecommunications regarding a draft bill and three draft Royal Decrees regarding mobile networks (2019)
65. OFCA: Guideline GN-13/2019. Guidelines for Submission of Applications for Assignment of Shared Spectrum in the 26 GHz and 28 GHz Bands (2019)
66. MIC: Announcement of the Local 5G Implementation guidelines, ICT X Japan (2019)
67. ITU-R: Resolution 229 (Rev WRC-19), Use of the bands 5150–5250 MHz, 5250–5350 MHz and 5470–5725 MHz by the mobile service for the implementation of wireless access systems including radio local area networks (2019)
68. FCC: CFR Title 47, Part 15, Subpart E. Unlicensed National Information Infrastructure Devices (2020)
69. FCC: NPRM 18-147. Unlicensed Use of the 6 GHz Band (2018)
70. ECC Report 302: Sharing and compatibility studies related to Wireless Access Systems including Radio Local Area Networks in the frequency band 5925–6425 MHz (2019)
71. Ofcom: Improving spectrum access for Wi-Fi Spectrum use in the 5 and 6 GHz bands, Consultation (2020)
72. ACMA: Future use of the 26 GHz band-Planning decisions and preliminary views (2019)
73. FCC: CFR Title 47, Part 15, §15.255, operation in 57–71 GHz (2020)
74. European Commission Decision 2013/752/EU of 11 December 2013 (amending Decision 2006/771/EC on harmonisation of the radio spectrum for use by short-range devices and repealing Decision 2005/928/EC) (2013)
75. Amit, R., Han, X.: Value creation through novel resource configurations in a digitally enabled world. Strateg. Entrep. J. **11**, 228–242 (2017)

Moving from 5G in Verticals to Sustainable 6G: Business, Regulatory and Technical Research Prospects

Marja Matinmikko-Blue[1]([✉]) [ID], Seppo Yrjölä[1,2] [ID], and Petri Ahokangas[3] [ID]

[1] Centre for Wireless Communications, University of Oulu, Oulu, Finland
marja.matinmikko@oulu.fi
[2] Nokia, Oulu, Finland
[3] Oulu Business School, Martti Ahtisaari Institute, University of Oulu, Oulu, Finland

Abstract. Mobile communication research is increasingly addressing the use of 5G in verticals, which has led to the emergence of local and often private 5G networks. At the same time, research on 6G has started, with a bold goal of building a strong linkage between 6G and the United Nations Sustainable Development Goals (UN SDGs). Both of these developments call for a highly multi-disciplinary approach covering the inter-related perspectives of business, regulation and technology. This paper summarizes recent advances in using 5G to serve vertical sectors' needs and describes a path towards sustainable 6G considering business, regulation and technology viewpoints. By focusing on key trends, the research summarizes four alternative scenarios for the futures business of 6G and considers related regulatory and technology aspects. Our findings highlight the importance of understanding the complex relations of business, regulation and technology perspectives and the role of ecosystems in both 5G in verticals and ultimately in the development of sustainable 6G to bring together stakeholders to solve long-term sustainability problems.

Keywords: Business strategy · Regulation · Scenario planning · Sustainability · 5G · 6G

1 Introduction

5G deployments are underway in the global scale with the first applications focusing on offering high capacity mobile broadband services. The promise of 5G to boost the digitalization of various vertical industries is gradually gaining increasing attention and the emergence of local 5G networks (Matinmikko et al. 2017, 2018) is starting to take place in some countries. Local 5G networks allow different stakeholders to use their own local connectivity platforms without having to rely on mobile network operators. These developments are occurring in complex multi-stakeholder ecosystems where regulatory, business, and technical perspectives are highly intertwined. The emergence of using 5G in the various verticals brings together the ICT sector and the vertical sector in question

G. Caso et al. (Eds.): CrownCom 2020, LNICST 374, pp. 176–191, 2021.
https://doi.org/10.1007/978-3-030-73423-7_13

with their own structures and rules for operations, calling for an ecosystem-level focus. Especially, the availability of spectrum for local networks fully depends on the country of operations, emphasizing the importance of regulatory decisions.

At the same time, research on the sixth generation (6G) of mobile communication networks has started globally aiming at first deployments in the 2030s. The first 6G White Paper published in 2019 presented a joint 6G research vision as a group work of 70 experts globally (Latva-aho and Leppänen 2019). The paper depicted the future 6G networks as an intelligent system of systems that combines the communication services with a set of other services including imaging, sensing, and locationing services, opening a myriad of new application areas. A set of continuation 6G White Papers published in 2020 (6G Flagship White Papers 2020) prepared in collaboration with 250 international experts went more into details and presented e.g. alternative future scenarios for the business of 6G (Yrjölä et al. 2020a), and developed a tight linking between 6G and the United Nations Sustainable Development Goals (UN SDGs) (Matinmikko-Blue et al. 2020a). Some of the developed future 6G business scenarios have taken sustainability as the starting point, stressing that the whole development of the future mobile communication networks should aim at helping society at large in its attempts to meet the sustainable development goals (Latva-aho and Leppänen 2019; Yrjölä et al. 2020a, b; Matinmikko-Blue et al. 2020a).

To make sense of moving from 5G in verticals towards 6G, we must envision future 6G systems targeting 2030 holistically from the perspective of the interaction between business, regulation and technology perspectives in envisioning future research prospects. The alternative futures of 6G will be shaped by growing societal requirements like inclusivity, sustainability, resilience, and transparency – a highly complex area that will call for major changes in industrialized societies in the long run, see (Latva-aho and Leppänen 2019; Matinmikko-Blue et al. 2020a). The business perspective specifically needs to consider sustainability (Kuhlman and Farrington 2010; Evans et al. 2017) in a way that combines the economic (e.g., profit, business stability, financial resilience, viability), societal (e.g., individuals', communities', regulative values) and environmental (e.g., renewables, low emissions, low waste, biodiversity, pollution prevention) perspectives. As an emerging field, 6G business scenarios and strategies have not been widely discussed in the literature to date. However, vision papers on future communication needs, enabling technologies, the role of artificial intelligence (AI), and emerging applications have recently been published (Viswanathan and Mogensen 2020; Saad et al. 2019; Letaief et al. 2019). Furthermore, discussion has latterly expanded to 6G indicators of value and performance (Ziegler and Yrjölä 2020), the role of regulation and spectrum sharing (Matinmikko-Blue et al. 2020a), the antecedents of multi-sided transactional platforms (Yrjölä 2020), antecedents of the 6G ecosystem (Ahokangas et al. 2020a) and the exploratory scenarios of 6G business (Yrjölä et al. 2020a).

Building on the above discussion, this paper provides an overview of 5G in verticals towards sustainable 6G from business, regulation and technology perspectives and presents related research prospects. The paper summarizes future scenarios for sustainable 6G business strategies in the timeframe 2030–2035, originally documented in (Yrjölä et al. 2020a), and related strategic options. The rest of this paper is organized as follows. Section 2 summarizes the state of the art of 5G in verticals from business, regulation and technology perspectives. Section 3 presents an overview of sustainable 6G. Future business scenarios for sustainable 6G and related strategic options are presented in Sect. 4. Finally, future outlook and conclusions are provided in Sect. 5.

2   State of the Art of 5G in Verticals

5G has been set high in national agendas to speed up digitalization of various sectors of society in many countries. This chapter presents recent developments in the use of 5G networks to serve the needs of different vertical sectors, such as industry, energy, and health, and their public sector counterparts, from the interrelated business, technology and regulation perspectives.

2.1   Business Perspective

Business perspective plays an important role in understanding the opportunities that a new technology can offer. The identification of the opportunity space for 5G business in verticals requires discussing four inter-related key themes: 1) the convergence of connectivity and data platforms and related ecosystems, 2) enablers, barriers and limitations to scalability and replicability of 5G solutions and business models, 3) legitimation of the new roles and business models within the verticals, and the 4) economic, societal and environmental sustainability of 5G solutions and business models. As vertical 5G networks are often considered as local networks, the platform-based business models utilized by different stakeholders face several challenges related to the afore-mentioned themes.

Mobile communication networks have for long been seen as platforms (Pujol et al. 2016) or ecosystems (Basole and Karla 2011). However, with the deployment of 5G networks, the mobile connectivity platforms operated by mobile network operators (MNOs) are increasingly becoming converged with the data platforms of various cloud service providers, giving rise to novel kinds of platform ecosystems. In industry verticals also the Industry 4.0 platforms as a specific type of data platforms play an important role. Extant literature identifies centralized, hybrid and fragmented types of converged connectivity and data platforms for industry verticals (Ahokangas et al. 2020b). In this kind of vertical context, a key feature of the converged platforms is the degree of openness achieved for different stakeholders of the ecosystem. Related to openness, the complexity, complementarity and interdependence of the converged connectivity and data platforms can be clarified by looking at the various components, interfaces, data and algorithms utilized in these platforms (Yrjölä et al. 2019) in connection to the connectivity (5G or other), content (e.g., information or data), context (location- or use-case specific data) or commerce (offering made available via a platform) business models utilized (Iivari et al. 2020). The vertical business model for local 5G operators presented by (Ahokangas et al. 2019) builds specifically on the idea to provide tailored end-to-end services in restricted geographical areas, such as industry sites, to the users locally. Vertical business models form a vertically structured ecosystem around the activity. The presented oblique business model and corresponding oblique ecosystem in turn builds on mass-tailored end-to-end services with stricter requirement for segmentation (Ahokangas et al. 2019).

The different types of converged connectivity and data platforms and the business models identified for them have varying potential for scalability and replicability. A scalable business model is agile and provides exponentially increasing returns to scale in terms of growth from additional resources applied (Nielsen and Lund 2018), whereas a replicable business model can be copied to several markets simultaneously with minimum variations (Aspara et al. 2010). For a firm running a vertical business model, scalability is based on the firm's capability to understand customer-specific needs and fulfill them, but limited on the size of the cases, their volume and timeline. For a firm running an oblique business model, scalability is based on the volume of unmet local needs and limited by access and availability of local infrastructures needed for providing the service (Ahokangas et al. 2019).

Within converged connectivity and data platform ecosystems, different stakeholders have varying roles and can act as service providers. This raises the issue of legitimacy, meaning that the activities of the stakeholder providing the service is legal and fits with the institutionalized practices within the industry in question (Marano et al. 2020). Achieving legitimacy for local vertical-specific 5G services and service providers through the deployment of local 5G networks is, however, an open question in many countries. Indeed, disruptive innovations such as 5G have been found to cause regulatory, incumbent and social "pushbacks" and they can be expected also for vertical 5G services, as legitimacy is a precondition for successful value creation and capture on a technology (Biloslavo et al. 2020).

The above discussion points out several challenges for reaching sustainable business models in 5G verticals. "A business model for sustainability helps describing, analyzing, managing and communicating (i) a company's sustainable value proposition to its customers, and all other stakeholders, (ii) how it creates and delivers this value, (iii) and how it captures economic value while maintaining or regenerating natural, social, and economic capital beyond its organizational boundaries" (Schaltegger et al. 2016, p. 6). Building vertical 5G business opportunities calls thus for filling in the requirements of scalability, replicability, and sustainability in a legitimate way in a platform ecosystem comprising connectivity and data services.

2.2 Regulation Perspective

The serving of the different verticals with 5G networks is not only addressed by the current MNOs but increasing attention is being paid to local and often private 5G networks (Matinmikko et al. 2017, 2018) that can be operated independent of the MNOs. Their emergence is highly dependent on the regulations that govern both the electronic communications market as well as the specific verticals, leading to a complex environment to operate. Regulations at national, regional and international levels define the operational conditions and there is wide variation between the national approaches but also some level of harmonization such as on the spectrum for 5G.

Prior work on regulatory developments on local 5G networks (Matinmikko et al. 2018; Vuojala et al. 2019; Lemstra 2018; Ahokangas et al. 2020b) have considered access regulation, pricing regulation, competition regulation, privacy and data protection, and authorization of networks and services. Especially, the authorization of networks and services defining the ways how rights to use radio frequencies are granted is critical for the establishment of local private 5G networks. Without the timely availability of sufficient amount of spectrum suitable for operations in the given environment, it is not possible to deploy the local networks. Specific spectrum options for local 5G networks are analyzed in detail in (Vuojala et al. 2019) including unlicensed access, secondary licensing, spectrum trading/leasing, virtual network or local licensing. Local licensing has emerged as a new spectrum access model in 5G to allow different stakeholders to deploy local networks in addition to the MNOs. A study on the recent 5G spectrum awards decisions in the 3.5 GHz band presented in (Matinmikko-Blue et al. 2019) shows that there is a big divergence in the spectrum awards by different countries taken by the regulators globally.

5G regulatory situation in Europe is discussed in (Lemstra 2018) where two contrasting scenarios for the future telecommunication market are presented including evolutionary and revolutionary scenarios. Evolutionary scenario continues the MNO market dominance which is likely to occur under the current European regulatory framework. The revolutionary scenario introduces new virtual MNOs that serve specific industry sectors which calls for additional policy and regulatory measures. The mobile communication market is in a turning point with the emergence of locally operated 5G networks by different stakeholders, especially aiming at serving the verticals' specialized local needs.

2.3 Technology Perspective

Previous generation mobile technologies have been largely deployed by national (or multi-national) incumbent MNOs for public use, given the high levels of investments required for the infrastructure, and to acquire exclusive radio spectrum. Furthermore, management and operational costs of the networks have been significant, and mobile technologies have required large and complex system integration from global infrastructure vendors with specialized capabilities. In addition to improved performance characteristics in capacity, speed and latency, novel 5G architecture is bringing additional flexibility for traditional MNOs as well as local operators in system deployments. Key technologies expected to transform 5G for verticals include localization and decomposition of network functions, software defined networking and network virtualization among others (Morgado et al. 2018).

A critical aspect of the local private industrial 5G networks is the ability to create customized network slices, where instances of virtual network resources and applications can be delivered to a new breed of services tailored to specific customer or tenant needs with service level agreed performance on demand. Furthermore, the software-based network architecture enables efficient sharing of common network infrastructure and resource by different tenants. Abstracting the slice functionality through open interfaces exposure to third party service provisioning enables service-dominant model for the connectivity and underlying network resources, e.g., computing, data and intelligence. The evolution towards the cloud-native infrastructure abstraction both on core and radio access empowers technology vendors and service providers to deploy and operate flexible and portable processes and applications in dynamic multi-vendor cloud environments. The cloud embedded in the edge of the network provides tools for optimized performance and economics for both the virtualized network functions and any other performance critical enterprise or vertical service and can become a control point of the local connectivity and intelligence. Edge cloud use cases considered in 5G are e.g., cloud radio access network (Open RAN, Virtual RAN), edge security, network and service automation enhancing the network itself, and industrial automation, massive scale Internet of Things (IoT), and augmented intelligence with augmented reality (AR)/virtual reality (VR). Another critical aspect is the spectrum. Operations in higher carrier frequencies represent a challenge in terms of deployment. The availability of suitable spectrum for serving the verticals cannot be based on dedicated spectrum paradigm but requires sharing in different domains.

Figure 1 summarizes the presented business, regulation and technology perspectives for 5G in verticals.

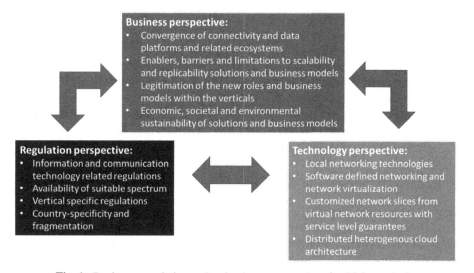

Fig. 1. Business, regulation and technology perspectives for 5G in verticals.

3   Towards Sustainable 6G

In parallel with the on-going development and deployment of 5G in verticals, research on the next generation, namely 6G, systems has already started in different parts of the world, see (Latva-aho and Leppänen 2019) and (6G Flagship White Papers 2020). The research on 6G (Latva-aho and Leppänen 2019; Matinmikko-Blue et al. 2020a) has identified sustainability stemming from the UN SDGs as the starting point, and it needs to address the technical, business and regulation perspectives, which are discussed next.

3.1   Role of UN SDGs in 6G

Future 6G networks are aiming at first deployments around the year 2030 which is also the target year for the achievement of the UN SDGs. While 6G communications is expected to boost global growth and productivity, create new business models and transform many aspects of society, its linking with the UN SDGs needs to be clearly formulated. The starting point of 6G research vision presented in (Latva-aho and Leppänen 2019) is that the development of 6G should be fully aligned with the UN SDGs (United Nations 2018). In a follow-up white paper, Matinmikko-Blue et al. (2020a) have developed a novel linking between 6G and the UN SDGs through the indicators of the UN SDG framework.

In (Matinmikko-Blue et al. 2020a) a three-fold role is foreseen for 6G as 1) provider of services to help reaching the UN SDGs, 2) enabler of measuring tools for data collection to help with the reporting of indicators, and 3) reinforcer of a new ecosystem to be developed in line with the UN SDGs. The white paper further details the linking between 6G and UN SDGs trough the existing indicators of the UN SDG framework where only 7 out of the 231 individual indicators are identified as being related to ICT. In reality, the ICT sector can influence many of the indicators, if not all. The white paper (Matinmikko-Blue et al. 2020a) analyses what 6G can do to contribute to the different UN targets within the SDG framework via the existing UN SDG indicators. The white paper proceeds to stating the need for a new set of indicators for 6G, characterizing the three-fold role of 6G. Additionally, a preliminary action plan is presented, calling for research and educational organizations, governments, standards developers, users, MNOs, network equipment manufacturers, application and service providers and verticals to think out-of-the box and create new technology solutions and collaborative business models to develop new operational models that support the achievement of the SDGs which may need changes to the existing regulations.

3.2 Business, Regulation and Technology Perspectives

The discussion on 5G business perspective for deployment in verticals presented in Sect. 2.1 proposed to focus on business models as a way of thinking future 6G ecosystem stakeholders' choices regarding opportunities, value-add and capabilities, and their expected consequences as scalability, replicability, and sustainability. With the right business choices, opportunities will be identified related to novel and unmet needs, new types of customer and service provider, as well as the interfacing of humans with machines in 6G. New value-add is seen to come from real-time and trustworthy communications, the use of local data and intelligence, and the commoditization of 6G resources as its competitive advantages, including extreme capacity and security, transaction and innovation platformization, and ubiquitous access. The expected business consequences of scalability may be related to the long tail of services, dataflow architecture, automation, and open collaboration between stakeholders; in terms of replicability, to deliberately design modularity and complementarity within platforms; and in terms of sustainability, to empower users and communities, and the utilization of sharing economic mechanisms in the markets.

Overall, governments and industries are under high pressure from the sustainability targets arising from the UN SDGs to renew their operations and the achievement of the goals provides new business opportunities especially for ICT solutions. These data and connectivity solutions can significantly contribute to industries to improve their resource efficiency and reduce waste but the solutions themselves need to be developed in alignment with the sustainability goals as well.

Digital convergence across industries and multi-level 6G platforms and ecosystems are creating a complex strategic environment that can lead to incomparable and distinct opportunities, as well as emergent problems. The regulations governing the use of future telecommunication systems and the relevant industry specific regulations together create a complex environment, especially around the use of data and connectivity platforms for different purposes. In particular, unanswered questions remain about ecosystemic business models in the context of sustainability. According to our recent findings (Yrjölä et al. 2020a, b), business ecosystems that aim to bring together stakeholders to solve systemic sustainability problems will require open ecosystem-focused value configuration and decentralized power configuration, where traditional stakeholder roles change, and new roles emerge. The focus needs to be on the long tail of specialized user requirements that crosses a variety of industries where related needs can be met with different resource configurations.

Spectrum continues to be the key resource for 6G systems as for any wireless networks throughout the times, and the availability of suitable spectrum continues to be significantly restricted due to the existing incumbent spectrum usage, see (Matinmikko-Blue et al. 2020b). Spectrum availability is a good example of the complex relations of business, regulation and technology perspectives. The availability of spectrum is a regulation decision, which defines the business opportunities and yet is restricted with technical aspects. Potential operations of future 6G systems in the new higher frequency bands at upper millimeterwaves (mmW) and terahertz (THz) regions pose significant technical, regulatory and deployment related challenges. Therefore, future 6G is not restricted only to higher frequency bands but can also be used in the existing bands for mobile communications. What are the economically feasible operational models, how to protect existing incumbent users of the feasible bands and how to implement THz radio links continue to be open topics for 6G.

The technology vision work in the global scale for systems towards 2030 and beyond has started at the International Telecommunication Union Radiocommunication sector (ITU-R) with the development of a report on future technology trends. The need for new indicators to characterize the performance of future 6G networks is evident (Latva-aho and Leppänen 2019; Matinmikko-Blue et al. 2020a; Pouttu et al. 2020), especially for defining and measuring resource efficiency and particularly energy efficiency. Also, the network architecture of 6G needs to be re-thought from prior generations of networks, see (Taleb et al. 2020). Figure 2 provides a summary emphasizing the need to develop sustainable 6G in line with the UN SDGs from business, regulation and technology perspectives.

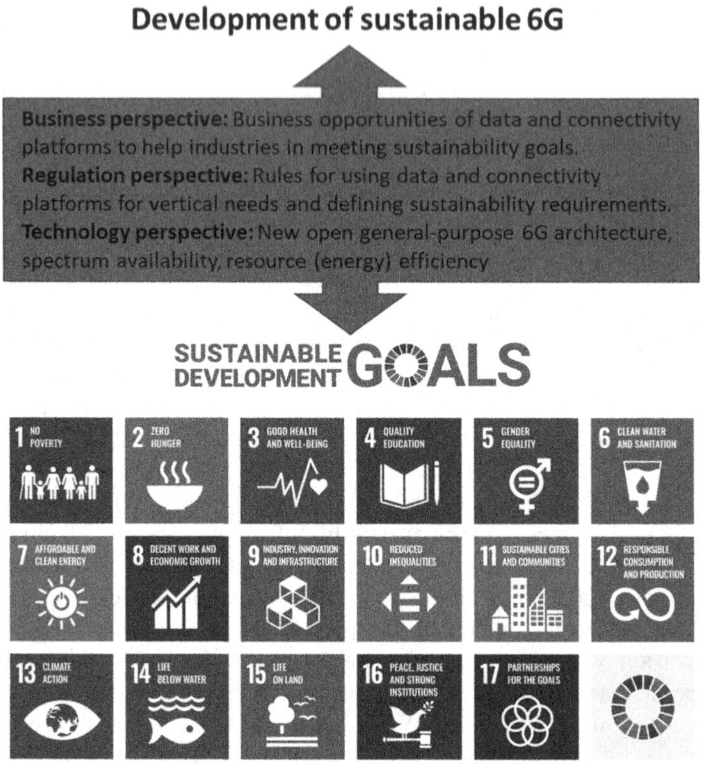

Development of sustainable 6G

Business perspective: Business opportunities of data and connectivity platforms to help industries in meeting sustainability goals.
Regulation perspective: Rules for using data and connectivity platforms for vertical needs and defining sustainability requirements.
Technology perspective: New open general-purpose 6G architecture, spectrum availability, resource (energy) efficiency

Fig. 2. Business, regulation and technology perspectives for sustainable development of 6G.

4   Business Scenarios and Strategic Options for 6G

Next, we proceed to new business scenarios developed for 6G and related strategic options developed through a set of virtual future-oriented white paper expert group workshops organized by 6G Flagship at the University of Oulu in 2020 and documented and analyzed in (Yrjölä et al. 2020a, b).

4.1   Methodology

The alternative scenarios for the future business of 6G summarized in this paper were created using anticipatory action learning (AAL) research method (Stevenson 2012) within 6G Flagship's white paper preparation (6G Flagship 2020). The process involved a series of online workshops in January–April 2020 where a group of experts from research, standardization and development, telecommunication industry, government, and verticals joined to collaboratively create future business scenarios for 6G.

First, the key change drivers for future 6G business were identified resulting in 153 forces (Yrjölä et al. 2020a). Using these drivers, a set of dimensions and endpoints were selected to form the basis for the scenario development as shown in Fig. 3. Value creation and value configuration were selected as the main business dimensions with different end points emphasizing closed and open alternatives.

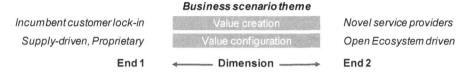

Fig. 3. Selected business scenario logic and dimensions.

We also used a simple rules strategy framework presented in (Eisenhardt and Sull 2001), which is a strategic management tool to develop strategies around identified business opportunities and describing the main processes. It provides a highly practical approach with guidelines in the following six rule categories introduced in (Eisenhardt and Sull 2001) and applied in the mobile communication market in (Ahokangas et al. 2013): 1) Nature of opportunity rules, 2) How to conduct business and processes in a unique way, 3) Boundary rules to decide, which opportunities to pursue, 4) Priority rules to identify and rank the opportunities, 5) Timing rules to synchronize emerging opportunities and other parts of the company, and 6) Exit rules to selecting things to be ended.

Next, we introduce the four developed business scenarios using the dimensions of Fig. 3 including Sustainable edge, Telco brokers, MNO6.0 and Over-the-top, as summarized in Fig. 4, and presented in (Yrjölä et al. 2020a). We also briefly summarize strategies as simple rules that were created for the most plausible MNO6.0 scenario and the most preferred Sustainable edge scenario.

4.2   6G Business Scenarios

A set of business scenarios were developed in 6G Flagship's white paper process in 2020, documented in (Yrjölä et al. 2020a) and summarized in the follows. Figure 4 summarizes the developed four business scenarios following the scenario logic of Fig. 3.

In the first scenario, the Sustainable Edge Value Creation, scenario in the upper-right corner of Fig. 4, the value creation is customer attraction-driven, and the value configuration is open ecosystem-focused. This scenario is built on decentralized open value configuration and ecosystem-driven business models where novel stakeholders take over customer ownership and networks. Changing stakeholder roles include web-scales, over the top (OTT) companies and device vendors being responsible for business to consumer (B2C) customers and local private cloud native networks serve business to business (B2B) customers. The role of traditional MNOs has changed into a wholesale connectivity service provider. Open source principles have become widely spread leading to technology and innovation ownership beyond traditional technology providers through open application programming interfaces (API) and novel resource brokerage. This scenario includes new stakeholder roles also in the form of local communities and special interest groups operating various edge resources in specific locations, such as

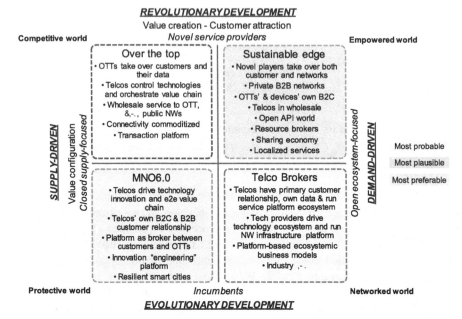

Fig. 4. Summary of developed 6G business scenarios.

campuses and remote areas to promote local innovation. New applications come with 6G technology that act as digital value platforms expanding our experiences towards digital computer-generated virtual worlds. The current focus on global-scale solutions changes towards local solutions that balance local demand with local supply and support circular economies. Especially the manufacturing vertical will move towards local decentralized manufacturing supporting a new crowdsourcing-based production ecosystem.

In the second scenario, the Telco Broker Value Creation by Incumbents and Open Ecosystem Value Configurations scenario shown in lower-right corner of Fig. 4, the main drivers for value creation remain the existing MNOs while value configuration is based on open ecosystem-focus. The MNOs are in charge of customer relationships and use service platform ecosystem to capture value. Technology providers' role is to develop the required technologies and provide network infrastructure via platform-based ecosystemic business models. Innovation ecosystem is broadened by the decoupling of technology platforms. Industry 5.0 (I5.0) has emerged as a key vertical for collaborative human machine interaction with robotization across services and industries. Real-time data and high level of digital automation allow the industries to focus on servitization of products. The speed of operations gets more and more rapid within the increasingly reprogrammable and reconfigurable world where design focus gets more and more short-term.

In the third scenario, the MNO6.0 Value Creation scenario shown in the lower-left corner of Fig. 4, value creation is driven by the incumbent MNOs, and value configuration is closed supply-focused. The role of MNOs is strong, and they drive technological innovation and own the customer relationships. The existing dominant MNO market position with strong customer base acts as the opportunity for businesses is and the focus is on how to cost-efficiently increase the capacity to meet the growing demand. Technology developments on dynamic networks slicing allowing increasing flexibility, shorter time-to-market, and cost optimization. With the MNO market dominance, the use of 6G in verticals is heavily dependent on MNOs' business decisions. Key technology developments in the form of automated network slicing and operations in higher frequency bands and new machine learning inspired tools will be used to optimize network operations in a predictive manner allowing new applications. These networks will have been assembled with a public-private-partnership funding model, with a view to resiliency and sustainability.

In the fourth scenario, the Over-the-Top Value Creation scenario shown in upper-left corner of Fig. 4, value creation is customer attraction- and lock-in-driven, and value configuration is closed supply-focused. The MNO dominance is replaced by OTTs that have taken over the customer relationships with the help of their access to customer data. The role of operators is to control the standardized and commoditized connectivity technologies and manage the value chains. The role of edge computing is to act as a new control point for serving of the verticals. Networks are programmable and make use of digital twins that represent replicas of complex physical systems to help in optimizing these systems. The ecosystem gets increasingly complicated with different resources and assets needed to meet the versatile needs are brought together by a set of stakeholders including physical infrastructure providers, equipment providers, and data providers under a complex regulatory framework defined by policymakers. Countries with more permitting rules act as resource pools and offer cheap labor, natural resources, and data.

The four developed scenarios were then assessed in terms of their probability, plausibility and preferability. Both the most probable scenario was the Over-the-top scenario while the most plausible scenario was the MNO6.0 scenario. The most preferable scenarios was the Sustainable edge scenario that can be seen to take a bold step towards achievement of the UN SDGS, representing revolutionary and demand-driven transformations.

The developed business scenarios for 6G indicate that from economic perspective, user experiences will be increasingly local and customized, delivered by local supply models supporting spatial circular economies. New societal service delivery models will appear through community-driven networks and public private partnerships and the role of 6G will be substantial in vertical industries. The developed scenarios revealed interesting societal observations including increasing tensions between competitive, protective, networked and empowered worldviews. The role of power configurations keeps increasing and may shift from a multi-polarized world to a poly-nodal world.

The pressure on companies and governments to meet the UN SDGs is evident in the business scenarios for 6G and the role of 6G as a provider of services towards environmental impact will be important. 6G with a set of new technologies will help in the monitoring and steering of circular economy to promote a truly sustainable data economy. The developed scenarios also show that 6G development faces privacy and security issues related to business and regulation including different aims of governance either stemming from governmental, company or end user perspectives. There the ecosystem-level configurations related to users, decentralized and community-driven business models and platforms and related user empowerment become increasingly important to support the role of local 6G services.

4.3   Strategic Options for 6G as Simple Rules

Next, we summarize the developed strategic options for selected two scenarios using the simple rules framework from (Eisenhardt and Sull 2001) that was applied to characterize MNOs' strategic choices in (Ahokangas et al. 2013). For the most plausible MNO 6.0 scenario, the baseline for building the simple rules is in the use of MNOs' wide existing customer base that has growing capacity needs through investments to strengthen customer lock-in and dominant market position in connectivity, enhanced with customer data and holding on to spectrum. The goal is to maintain dominant market position through gaining access to a new wideband spectrum. Automation of network operations and the ability to dynamically create large numbers of networks slices on-demand will help to increase flexibility, shorten time-to-market, and optimize costs. Resources and services will be traded in automated marketplaces. The MNOs could become a wholesale platform provider for other operators which would further strengthen their market position. Regulations plays a key role in maintaining the MNO market dominance which calls for close contact with the regulator. In the MNO6.0 scenario, the MNOs would never give up on their spectrum and customer data.

For the most preferred Sustainable Edge Scenario, the simple rules are built on the use of new, local, and specialized demand, challenging incumbent MNOs in narrow business segments specializing in governmental, municipal, vertical, or enterprise customers and vertical differentiation with increasing requirements for sustainability in specific industry segments like education, healthcare, and manufacturing. These challenger operators think and act locally, close to the customer and promote resource sharing in different format such as spectrum and virtualized cloud infrastructures. Sustainability requirements in verticals are a major business opportunity through providing vertical differentiation in specific segments like education, healthcare, manufacturing, energy, and media and entertainment. The sustainable edge service provider supports circular economy and promotes sharing economy principles in network deployment. These locally operated networks have opportunities to scale up from local operations to a multi-locality business. Local and private networks provide several benefits in terms of security and data control, separation from public networks, access to high-quality services in specific locations, increased flexibility, scalability and customization, and trustworthy reliabilities and latencies. Furthermore, networks can be deployed as standalone sub-networks or integrated with MNO networks. This requires the establishment of multi-sided platforms-based regulations to govern privacy and security of users.

5   Future Outlook and Conclusions

Mobile communication research is increasingly addressing the use of 5G in verticals, which has led to the emergence of local 5G network deployment models. Research on 6G has also started, with a bold goal of building a strong linkage with the United Nations

Sustainable Development Goals (UN SDGs). These developments call for a highly multi-disciplinary approach covering business, regulation and technology perspectives and our research is addressing these interrelated themes. This paper has provided an overview of the recent developments in 5G in verticals towards the development of sustainable 6G. We have highlighted the importance of the triangle of business – regulation – technology perspectives in the development of new wireless technologies and their deployments and summarized the advancements with a focus on local 5G networks for serving the verticals' needs towards meeting the sustainable development goals.

From the business perspective, a business model for sustainability can help in describing, analyzing, managing and communicating 1) a company's sustainable value proposition to its customers, and other stakeholders, 2) how it creates and delivers this value, 3) and how it captures economic value while maintaining or regenerating natural, social, and economic capital beyond its organizational boundaries. The development of new vertical-specific 5G business opportunities calls for filling in the requirements of scalability, replicability, and sustainability in a legitimate way in a platform ecosystem of connectivity and data services. Digital convergence across industries and multi-level 6G platforms and ecosystems will create a complex environment where ecosystemic business models for sustainability and the evolution of related regulations become important. Business ecosystems that aim to bring together stakeholders to solve systemic sustainability problems will require open ecosystem-focused value configuration and decentralized power configuration, focusing on the long tail of specialized user requirements that crosses a variety of industries. Future research prospects are particularly related to the new business ecosystems, ecosystemic business models and changing stakeholder roles that support sustainability.

From the regulation perspective, the serving of different verticals with 5G and future 6G networks introduces local and often private wireless networks to complement the current mobile network operators (MNOs). The regulatory environment for 5G in verticals is very complex encompassing rules from both the electronic communications market as well as specific verticals. Especially, the ways how rights to use radio frequencies are granted is critical for the establishment of local 5G and 6G networks. The divergence in spectrum awards between countries is increasing with 5G, directly influencing the business opportunities in those countries. There are research prospects in finding the best practices from the decisions by analyzing their impact.

For the technology perspective, 5G and future 6G architecture is expected to bring additional modularity and flexibility for traditional MNOs as well as for new local operators in system deployments. Key technologies to enable open general-purpose 6G architecture include distributed heterogenous cloud-native architecture, localization and decomposition of network functions, software defined networking and network virtualization, among others. A critical aspect for the local private industrial networks is their ability to create customized network slices that allow the delivery of services tailored to specific customer needs with service level agreed performance on demand. The availability of spectrum for serving the verticals and operations in higher carrier frequencies present a major technical deployment challenge. The availability of spectrum for serving the verticals on shared basis is important. New research prospects are especially in the

6G domain in order to find new indicators for 6G that take sustainability in to account as well as the new network architecture for 6G needs.

This study has identified a further need for foresight research that explores the inter-related business – regulation – technology perspectives in the context of 5G in verticals and on the road to sustainable 6G, with a special focus on how can 6G become a truly general-purpose technology instead of simply an enabling technology, to support countries and organizations in the journey towards the achievement of the UN SDGs. Especially, the verticals burdened by increasing requirements for sustainability will be in the key position in to realize the benefits of using the new technologies.

References

6G Flagship: White Papers. https://www.6gchannel.com/6g-white-papers/. Accessed 24 July 2020 (2020)

Ahokangas, P., Matinmikko, M., Yrjölä, S., Okkonen, H., Casey, T.: Simple rules for mobile network operators' strategic choices in future cognitive spectrum sharing networks. IEEE Wirel. Commun. **20**(2), 20–26 (2013)

Ahokangas, P., et al.: Business models for local 5G micro operators. IEEE Trans. Cogn. Commun. Netw. **5**(3), 730–740 (2019)

Ahokangas, P., Yrjölä, S., Matinmikko-Blue, M., Seppänen, V. Transformation towards 6G ecosystem. In: Proceedings of 2nd 6G Wireless Summit, Levi, Finland (2020a)

Ahokangas, P., Matinmikko-Blue, M., Yrjölä, S., Hämmäinen, H.: Future vertical 5G platform ecosystems: case study of a 5G enabled digitalized port stakeholders' new interactions and value configurations. In: Proceedings of International Telecommunications Society Online Conference, Gothenburg, Sweden (2020b)

Aspara, J., Hietanen, J., Tikkanen, H.: Business model innovation vs replication: financial performance implications of strategic emphases. J. Strateg. Mark. **18**(1), 39–56 (2010)

Basole, R.C., Karla, J.: On the evolution of mobile platform ecosystem structure and strategy. Bus. Inf. Syst. Eng. **3**(5), 313 (2011)

Biloslavo, R., Bagnoli, C., Massaro, M., Cosentino, A.: Business model transformation toward sustainability: the impact of legitimation. Manag. Decis. **58**, 1643–1662 (2020)

Eisenhardt, K.M., Sull, D.N.: Strategy as simple rules. Harvard Bus. Rev. **79**(1), 107–116 (2001)

Evans, S., et al.: Business model innovation for sustainability: towards a unified perspective for creation of sustainable business models. Bus. Strateg. Environ. **26**, 597–608 (2017)

Iivari, M., Ahokangas, P., Matinmikko-Blue. M., Yrjölä, S.: Opening closed business ecosystems boundaries with digital platforms: empirical case of a port. In: Ziouvelou, X., McGroarty, F. (eds.) Emerging Ecosystem-Centric Business Models for Sustainable Value. IGI Global (2020)

Kuhlman, T., Farrington, J.: What is sustainability? Sustainability **2**(11), 3436–3448 (2010)

Latva-aho, M., Leppänen, K. (eds.): Key Drivers and Research Challenges for 6G Ubiquitous Wireless Intelligence. 6G Research Visions 1, University of Oulu, Finland (2019)

Lemstra, W.: Leadership with 5G in Europe: two contrasting images of the future, with policy and regulatory implications. Telecommun. Policy **42**(8), 587–611 (2018)

Letaief, K.B., Chen, W., Shi, Y., Zhang, J., Zhang, Y.A.: The roadmap to 6G: AI empowered wireless networks. IEEE Commun. Mag. **57**(8), 84–90 (2019)

Marano, V., Tallman, S., Teegen, H.J.: The liability of disruption. Glob. Strategy J. **10**(1), 174–209 (2020)

Matinmikko, M., Latva-aho, M., Ahokangas, P., Yrjölä, S., Koivumäki, T.: Micro operators to boost local service delivery in 5G. Wirel. Pers. Commun. **95**(1), 69–82 (2017)

Matinmikko, M., Latva-aho, M., Ahokangas, P., Seppänen, V.: On regulations for 5G: micro licensing for locally operated networks. Telecommun. Policy **42**(8), 622–635 (2018)

Matinmikko-Blue, M., Yrjölä, S., Seppänen, V., Ahokangas, P., Hämmäinen, H., Latva-Aho, M.: Analysis of spectrum valuation elements for local 5G networks: case study of 3.5-GHz band. IEEE Trans. Cogn. Commun. Netw. **5**(3), 741–753 (2019)

Matinmikko-Blue, M., et al. (eds.): White Paper on 6G Drivers and the UN SDGs. 6G Research Visions 2, University of Oulu, Finland (2020a)

Matinmikko-Blue, M., Yrjölä, S., Ahokangas, P.: Spectrum management in the 6G era: role of regulations and spectrum sharing. In: Proceedings of 2nd 6G Wireless Summit, Levi, Finland (2020b)

Morgado, A., Huq, K.M.S., Mumtaz, S., Rodriguez, J.: A survey of 5G technologies: regulatory, standardization and industrial perspectives. Digital Commun. Netw. **4**(2), 87–97 (2018)

Nielsen, C., Lund, M.: Building scalable business models. MIT Sloan Manag. Rev. **59**(2), 65–69 (2018)

Pouttu, A. (ed.): 6G White Paper on Validation and Trials for Verticals towards 2030's. 6G Research Visions 4, University of Oulu, Finland (2020)

Pujol, F., Elayoubi, S.E., Markendahl, J., Salahaldin, L.: Mobile telecommunications ecosystem evolutions with 5G. Commun. Strateg. **102**, 109 (2016)

Saad, W., Bennis, M., Chen, M.: A vision of 6G wireless systems: applications, trends, technologies, and open research problems. IEEE Netw. **34**, 134–142 (2019)

Schaltegger, S., Hansen, E.G., Lüdeke-Freund, F.: Business models for sustainability: origins, present research, and future avenues. Organ. Environ. **29**(1), 3–10 (2016)

Stevenson, T.: Anticipatory action learning: conversations about the future. Futures **34**, 417–425 (2012)

Taleb, T., et al.: White Paper on 6G Networking. 6G Research Visions 6. University of Oulu, Finland (2020)

United Nations: Global indicator framework for the Sustainable Development Goals and targets of the 2030 Agenda for Sustainable Development (2018)

Viswanathan, H., Mogensen, P.E.: Communications in the 6G era. IEEE Access **8**, 57063–57074 (2020)

Vuojala, H., et al.: Spectrum access options for vertical network service providers in 5G. Telecommun. Policy **44**(4), (2019)

Yrjölä, S., Ahokangas, P., Matinmikko-Blue, M.: Novel platform ecosystem business models for future wireless communications services and networks. In: Proxceedings of NFF 2019, Vaasa, Finland (2019)

Yrjölä, S., Ahokangas, P., Matinmikko-Blue, M. (eds.): White Paper on Business of 6G. 6G Research Visions, No. 3. University of Oulu, Finland (2020a). http://urn.fi/urn:isbn:978952622 6767

Yrjölä, S.: How could blockchain transform 6G towards open ecosystemic business models? In: Proceedings of IEEE ICC 2020 Workshop on Blockchain for IoT and CPS, Dublin, Ireland (2020)

Yrjölä, S., Ahokangas, P., Matinmikko-Blue, M.: Sustainability as a challenge and driver for novel ecosystemic 6G business scenarios. Sustainability **12**(21), 8951 (2020b)

Ziegler, V., Yrjölä, S.: 6G indicators of value and performance. In: Proceedings of 2nd 6G Wireless Summit, Levi, Finland (2020)

Author Index

Printed in the United States
by Baker & Taylor Publisher Services